MW01678711

Magnetic Air Cars

The History of Magnetic Motors and Compressed Air Cars

To: Robert Intile

The future is now!
we look forward to
working with your
organization.

Steven A Leonard

Steven Leonard

Copyright Notice

ISBN Number: 978-0-615-26429-5

Author: Steve Leonard
sleonard@magneticaircars.com

Editor: Ramman Kenoun
rkenoun@magneticaircars.com

Magnetic Air Cars, Inc.
521 Charcot Ave., Suite 109
San Jose, CA 95131

Tel.: (408) 324-1366
Fax: (408) 324-1343

Table of Contents

Acknowledgements by the Author

There are many people to thank for making this book and Magnetic Air Cars possible. First, a great deal of gratitude is in order for my wife, Racheal. She is the beauty I feel behind the Vision.

Special thanks goes to William Bellou who is the owner of Times Media Inc. Times Media Inc. is one of the largest independent newspapers in Silicon Valley. He was one of the first to believe and contribute to the development of the Magnetic Air Car and Generator.

I would like to thank our patent attorney Richard Main and the engineering team of Paul Donovan, George Ristau, Eric Ristau, Howard Dumas, Elijah Dumas and Elisha Dumas. Their engineering contributions were instrumental in the development of the Magnetic Air Car and Generator.

Additionally, I would like to thank our family and friends for believing in my Vision and helping it grow into the Company it is today. Finally, last but certainly not least, I would like to thank our contributing investors, the talented investment bankers, engineers, graphic designers, programmers, artists, advisors, lawyers and accountants working with Magnetic Air Cars, Inc. Attracting and retaining the services of these individuals has been, and will continue to be, a primary key in keeping Magnetic Air Cars a growing and healthy Company.

The founders and writer's life experiences including the persons they have come in contact with and they have contributed knowingly and unknowingly combining a wealth of mind power to Magnetic Air Cars and this book.

Thank you everyone.

Dedications

This book is dedicated to Preston Tucker.

Preston Tucker had a dream of making a safe and reliable family automobile - with technological innovations that were considered radical at the time. As a result, the established car manufacturers considered his car a threat to their products. Tucker's car could be built for a fraction of the cost required for the mainstream carmakers to build one. His car also featured a wide array of extras like disc brakes, seat belts, fuel-injected engine in the rear, a padded dashboard, and a front windshield that popped out in a severe collision. These ideas were considered revolutionary at the time, and as Tucker began to make his car a reality, the powerful Detroit auto makers and the authorities in Washington, D.C. worked together to discredit him.

After World War II, entrepreneurs like Henry J. Kaiser and Preston Tucker saw an opportunity to enter the automobile market. The United States government was auctioning off surplus factories and giving preference to upstarts rather than large corporations which had benefitted from war production. Studebaker was the first with an all-new post-war model. However, Tucker took a different tack, designing a safety car with innovative features and modern styling. His specifications called for an air-cooled, flat 6-rear engine (not unlike the modern Porsche), disc brakes, fuel injection, the placement of all instruments on the steering wheel, and a padded dashboard.

Famed stylist Alex Tremulis, previously of Auburn/Cord/Duesenberg, was hired on December 24, 1946 and given just six days to finalize the design. On December 31, 1946, Tucker approved the design, dubbed the Tucker Torpedo. He also hired another firm to create an alternate body, but only the horizontal taillight bar from that specific model appeared on the final car.

A Tucker '48 Sedan design patent illustration

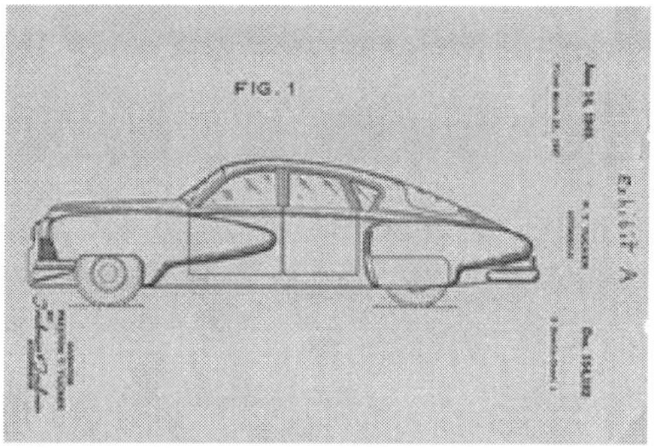

The mechanical components were highly innovative. The perimeter frame surrounded the vehicle for crash protection. The steering box was behind the front axle to protect the driver from injury in a front-end collision. Unfortunately, numerous Tucker innovations were dropped. Magnesium wheels, disc brakes, and a direct-drive torque converter transmission were all left on the drawing board.

However, the car's innovative engine continued on the production path for a while longer. It was a flat 6-cylinder with hemispherical combustion chambers, fuel injection, and overhead valves operated by oil pressure rather than a camshaft. These features would have been auto industry firsts in 1948, but as engine development proceeded, problems began to appear. The large 589 in^3 (9.7 L) engine functioned, but the 150 hp (112 kW) motor's valve train proved problematic. The engine and Cord transmission were mounted on a separate sub-frame, which could be lowered and removed in minutes by removing just six bolts. Tucker envisioned loaner engines being quickly traded in for service.

The final car was only 60" (1524 mm) tall, but was very roomy inside. It featured a directional third headlight, dubbed the "Cyclops Eye", for use in turns. It automatically lit up whenever the car was steered on an angle greater than 10 degrees. Tremulis' design was dubbed the most aerodynamic in the world, and though it still sported pre-war type fenders, it was startlingly modern. The mathematically-computed drag coefficient was only 0.27, although this figure was rounded up to 0.30 for the public. The world premiere of the much-hyped car was set for June 19, 1947. Over 3,000 people showed up for the launch, which included a train tour of the plant, and the unveiling in the main auditorium. The unveiling seemed to be doomed, however, as last-minute problems with the car cropped up. The suspension snapped and the car would not move. Tucker ad-libbed on stage for two hours while emergency repairs were carried out. It was finally pushed onto a turntable by hand, and the curtain was lifted to thunderous applause. Tucker was joined on stage by his family, with his daughter smashing a champagne bottle on the "Cyclops Eye" and soaking her father. Also appearing on stage were Tucker's engineers, still covered in grease from the last-minute repairs.

Tucker had promised 150 hp (112 kW), and his innovative 589 was not working, so another engine was sourced. The company first tried the Lycoming aircraft engine, but it would not fit in the car's rear engine compartment. The Franklin air-cooled helicopter's flat-6 engine did fit, however, so Tucker purchased four samples for $5,000 each. The company's engineers converted it to a water-cooled design, and its 166 hp (124 kW) capability pleased Tucker. He quickly acquired the Franklin Company to secure the engine source.

Another failed element of the prototypes was the Cord transmission, designed for front-engine/front-wheel drive use. It could not handle the power of the Franklin engine, so a new design was needed. The creator of the Buick Dynaflow transmission was called in, designing a special "Tuckermatic" transmission with only 27 parts, about 90 fewer than normal. Unfortunately, this transmission caused further issues, as the prototype lacked a reverse gear. A skeptical press reported that the car could not go backward, damaging its reputation.

Although well-funded, the company decided to raise more money in order to secure its future and credibility. $17,000,000 was raised in a stock issue, one of the first speculative IPOs. Another money maker was the Tucker Accessories Program. Future buyers could purchase accessories, like seat covers, the radio, and luggage, before their car was built. This brought an additional $2,000,000 into the company.

With the final design in place, Preston Tucker took the pre-production cars on the road to show them all across the country. The cars were an instant success, with crowds gathering wherever they stopped. One report says that

Tucker was pulled over by a police officer intent on getting a better look at the car.

Tucker Torpedo at Petersen Automotive Museum, Los Angeles, California

One of Tucker's most innovative business ideas caused trouble for the company. His Accessories Program raised funds by selling accessories before the car was even in production. This concept was investigated by the U.S. Securities and Exchange Commission and the United States Attorney, and led to an indictment of company executives. Despite all charges being eventually dropped, the negative publicity destroyed the company and halted production of the car. To counteract the bad press, Tucker once again took the cars on the road. He scheduled a two-week public test at the Indianapolis Motor Speedway with a few of the 37 cars that had been built. One car was rolled at 100 mph (160 km/h), and the driver walked away with only bruises. Public perception changed from outrage at Tucker's alleged misconduct to anger at the press and the government. Today, the '48 Sedan has fame far greater than would be expected from its modest production run. Of the 51 cars that were built (50 production and 1 prototype), 47 still exist, the majority in excellent condition. When the cars appear at auction, they command prices attained by only a few select makes. Car number 43 was recently sold for the record-setting price of $700,000. Several other cars are prominently featured in automotive museums. During the trouble that Tucker faced while trying to promote his car and get it into production, he claimed that the "Big Three" automakers were deliberately attempting to sabotage his efforts, through the influence of Detroit Senator Homer Ferguson, who is universally blamed for initiating the SEC's pursuit of Tucker's business.

At one point in his life, Mr. Tucker was admitted to a mental hospital.

This book is also dedicated to Dr. Nikola Tesla.

Tesla fired the imaginations of a perceptive few. Not only did Tesla teach by his accomplishments, but he taught by the inspiration of a marvelous imagination that refused to accept the permanence of what others perceived to be insuperable difficulties; an imagination the goals of which, in a number of instances, are still in the realms of speculation.

When the concept of the rotating magnetic field was revealed to Tesla, in a flash, he had seen the universe composed of a symphony of alternating currents with the harmonies played on a vast range of octaves. The 60-cycles-per-second AC was but a single note in a lower octave. In one of the higher octaves, at a frequency of billions of cycles per second, was visible light. He sensed that exploring this vast range of electrical vibration between his low-frequency alternating current and light waves would bring him closer to an understanding of the cosmic symphony.

Tesla was the first to build a marvelous induction motor for alternating current, based upon his discovery of the rotating magnetic field. Ninety-nine percent of genius was, "Knowing the things that would not work." Hence, Tesla approached each problem with an elaborate process of elimination. Tesla would proceed at once with the diligence of a bee to examine straw after straw until he found the object of his search. Other scientists like Thomas Edison believed that a little theory and calculation could have saved Tesla ninety percent of his labor. Edison and Tesla were singularly representative of different kinds of training, different methods, and different strains.

The U.S. Patent Office has 1,200 patents registered in the name of Nikola Tesla, and it is estimated that he could have patented an additional 1,000 or so from memory!

In 1931, under the financing of Pierce-Arrow and George Westinghouse, a 1931 Pierce-Arrow was selected to be tested at the factory grounds in Buffalo, N.Y. The standard internal combustion engine was removed and an 80 hp (1800 rpm) electric motor installed to the clutch and transmission. The AC motor measured 40" long and 30" in diameter and the power leads were left standing in the air, i.e., no external power source! At the appointed time, Nikola Tesla arrived from New York City and inspected the Pierce-Arrow automobile. He went to a local radio store and purchased a handful of tubes (12), wires and assorted resistors. With a box about 3 feet in length, Mr. Tesla got into the driver's seat and had its wires connected to the air-cooled, brushless motor. Two rods, ¼" in diameter, stuck out of the box and stated, 'We now have power.' He put the car into gear and it moved forward! This vehicle, powered by an AC motor, was driven to speeds of 90 mph and performed better than any internal combustion engine of its time! One week was spent testing the vehicle. Several newspapers in Buffalo reported this test. When asked where the power came from, Tesla replied, "From the ethers all around us." Several people suggested that Tesla was mad and somehow in league with sinister forces of the universe. He became incensed, removed his mysterious box from the vehicle, and returned to his laboratory in New York City. His secret died with him! It is speculated that Nikola Tesla was somehow able to harness the earth's magnetic field that encompasses our planet. Also, he was somehow able to draw tremendous amounts of power by cutting these lines of force or causing them to be multiplied together. The exact nature of his device remains a mystery, but it did function by powering the 80 hp AC motor in the Pierce-Arrow at speeds up to 90 mph, and no recharging was even necessary!

In 1969, Joseph R. Zubris took his 1961 Mercury and pulled out the internal combustion engine. He then installed an electric motor as a source of power. His unique wiring system cuts the energy drain at ignition to 75% of normal and doubles the electrical efficiency of the electric motor while in operation! The U.S. Patent Office issued him a patent (No. 3,809,978). Although he approached many with concerns regarding marketing, no one really seemed to be interested. In fact, his unique system is still not on the market. In the 1970's, an inventor used an Ev-Gray generator, which intensified battery current, the voltage being induced to the field coils by a simple programmer (sequencer). By allowing the motor to charge separate batteries as the device ran, phenomenally tiny currents were needed. The device was tested at the Crosby Research Institute of Beverly Hills, California. A 10 hp EMA motor ran for over a week (9 days) on four standard automobile batteries. The inventors estimated that a 50 hp electric motor could continue for 300 miles at 50 mph before needing a recharge. Dr. Keith E. Kenyon, the inventor from Van Nuys, California, discovered a discrepancy in the normal and widely-accepted laws relating to electric motor magnets. Dr. Kenyon demonstrated his invention for many scientists and engineers in 1976, but their reaction was astounding. Although they had admitted that Dr. Kenyon's device worked, they saw little or no practical application for it! Therefore, the ultimate source for our electrically-powered automobile would be an electric motor that required no outside source of power. Such a device seems impossible because it violates all scientific thought. However, it has been invented and H.R. Johnson was issued a patent (No. 4,151,431) on April 24, 1979 on such a device!

This new design, although originally suggested by Nikola Tesla in 1905, is a permanent magnet motor. Mr. Johnson had arranged a series of permanent magnets on the rotor, and a corresponding series – with different spacing – on the stator. One simply has to move the stator into position and rotation of the rotor begins immediately. His patent states, "The invention is directed to the method of utilizing the unpaired electron spins in Ferro magnetic and other materials as a source of magnetic fields for producing power without any electron flow, as occurs in normal conductors, and to permanent magnet motors for utilization of this method to produce a power source." In the practice of this invention, the unpaired motive power source, solely through the super-conducting characteristics of a permanent magnet and the magnetic flux created by the magnets, are controlled and concentrated to orient the magnetic forces generated in such a manner to do continuous work, such as the displacement of a rotor with respect to a stator. The timing and orientation of magnetic forces at the rotor and stator components produced by permanent magnets in order to produce a motor is accomplished with the proper geometrical relationship of these components.

Introduction

The desire to tell the truth presents complex issues. Several questions must be pondered before doing so. Are my recollections valid? Is my memory being selective? Does that selectivity constitute an untruth? Is my version of events really what happened, or from my singular point of view, only what I think happened? I sought other peoples' perspective on the events contained here. Then, I put these views together using other people's internal lenses, while striving to be fair and impartial.

This book is about a trend of the here and now, purified and intensified for dramatic effect and extended into the future. "If this goes on, this will happen next." A prediction is made and the future is revealed as if it already happened. A true artist, or seer, believes and knows what will happen because he felt the Spirit within him use his tongue and his hands, perhaps only once, but once is enough. I regret having used my imagination to reveal truth to you. I realize that the majority will believe that everything in this book is fabricated. There are two views that separate our understanding of reality. These two world views have always dramatically opposed each other from the beginning. Your reality and mine will probably never be the same.

Truth believes that anything you can imagine can be done.
A liar is someone who believes that anything you can imagine cannot be done.

Truth believes that nothing is impossible.
A liar believes that nothing is possible.

Truth believes in visions.
A liar isn't even familiar with the term.

Truth has faith, hope and dreams.
A liar has neither faith, hope, nor dreams.

Truth is called a liar because it believes.
A liar is so because it does not believe.

Truth can change the future by using its imagination.
A liar has no future, nor imagination.

Truth creates words.
A liar cannot even imagine.

Truth knows.
A liar does not.

The gatekeepers are the envious industries, the government, universities, scientists and critical journalists. They are the major cause of all travail on this earth. They have no imagination and do not believe in anything. They only believe in what can physically be observed in the present. They have no vision for our future.

Comment from David Waggoner…

"We have had people contact us over the years and tell us that… 'There is no such thing as free energy!' A statement like that is very sad and funny at the same time! Where have these people been? What about free energy from solar cell technology? What about free energy from atomic energy / nuclear power plants? The list goes on and on. Free energy is everywhere! The earth is loaded with free energy and acts as a motor or perpetual motion machine (so you could say). Hydrogen and oxygen atoms are free energy generators and are constantly being reused over and over without decay! Your body is basically a free energy - walking generator! A coil of wire induced by a DC pulse current will produce free energy in the form of back emf as well as other fields we are not able to discuss. I am not talking about perpetual motion machines I am talking about free energy!" – **David Waggoner**

Finally, I hope that when you're done reading this book, you'll be a bit different from what you were before you read it, that you will have changed a little, as if by having met a new face, experienced a new experience, and ventured into places you've never imagined.

Wisdom knows the lessons of history and can anticipate the future. She knows how to interpret what people say and how to solve problems. She knows the miracles that our Creator will perform and how the movements of history will develop.

A parable is a message of invitation handed down from the heavens in a veiled manner. It can either be a stumbling block or a challenge. A parable can either enlighten or become a blinding force.

Chapter 1

The Power Source

Starting Stage Subsystem 1

A battery-powered Starting State Subsystem 1. This part contains a 12v battery. The battery has three times the recharge rate of existing batteries and may last five times longer. The battery may be used to start our permanent magnet motor. The purpose of the 5-10 hp magnetic motor is to supply power to an air bearing compressor which will supply compressed air in real-time to the Center Stage Subsystem 2 for starting. The air bearing compressor may operate at up to 1000 ft^3 per minute. The duties of the permanent magnet motor may also be used to recharge the composite form grid battery. If the air bearing compressor fails to supply compressed air in real-time to the Center Stage Subsystem 2 for starting, a backup 120-gallon compressed air tank may also be used to supply compressed air in real-time to the Center Stage Subsystem 2 for starting. In the near future, our starting system may also incorporate a high density, stable and reliable tantalum-equivalent capacitor to start our permanent magnet motor.

A pulley system attached to the rear-wheel drive may also be connected to the permanent magnet motor to help assist in supplying power to an air bearing compressor which will supply compressed air in real-time to the Center Stage Subsystem 2 for starting. The mobility and motion of the vehicle may also be used to supply additional power to an air bearing compressor that supplies compressed air in real-time to the Center Stage Subsystem 2 for starting. Solar power may also be used to help assist in supplying power to an air bearing compressor that supplies compressed air in real-time to the Center Stage Subsystem 2 for starting. Subsystem 1 is in the business of producing highly compressed ambient air in real-time to the Center Stage Subsystem 2 for starting. As long as we can continue to supply highly compressed ambient air in real-time to the Center Stage Subsystem 2 for starting, we believe the Air

Flow Amplifier and the Air bearing turbocharger is capable of handling the load necessary for producing forward and backward mobility.

Chapter 2

Capacitors

A special class of capacitors, sometimes called super capacitors, has the potential to store large amounts of electricity in very small physical volumes. The most inconvenient property of electricity is that it is very difficult to store. This results in generation inefficiencies in which supply and demand are not matched. For example, let us say we are trying to "harvest" the abundant solar energy available in Southern California. The energy input (insulation) peaks at about 2 PM. However, the maximum demand for the power does not occur until 5PM. Storage of that power for 3 hours is very difficult. Batteries, pumped storage, and flywheels are very inefficient and waste up to 70% of the stored energy.

Capacitors may be able to handle a fifteen minute charge time with a 500 mile distance of travel. Currently, there is an electric car in Europe, with an 800 hp total. There may be a 200 hp motor in each wheel. The batteries are good for only 45 miles. However, it has an extremely small internal combustion engine that spins a generator and extends that distance. This would be a perfect candidate for driving a magnetic air motor.

Similarly, power plants, whether coal, oil or gas fired, are most efficient when operated at a constant output. Instead, it is necessary to design them to accept the maximum load. If there was an effective way to store the output, less fuel would be burned and the amount of copper used in transmission lines would be considerably less.

The capacitor technology developed by Mr. Nevin may well be an answer to this dilemma. Large amounts of electricity can be stored on non-planar conformal plates, allowing incomprehensible energy densities.

Parties wishing to learn more about this technology can contact Mr. Nevin at Nonplanar Nanostructures. Today, there is a method and apparatus for large-scale storage of electric potential in a zeolite-containing structure. The zeolite-containing structure is used as a substrate for at least one of the plates (conductive electrodes) of a charge storage device, or capacitor. Since the zeolite has a large surface area, the conductive electrodes made from and/or patterned upon that material will also have a large surface area, thereby giving the overall device a high capacity to store electric potential or charge. Within the structure lies an electrical storage device comprising a large surface area material (e.g. zeolite) substrate composed of at least one zeolite crystal having a surface, and a first electrically-conductive layer proximate to at least a portion of said surface, and a second electrically-conductive layer, with a dielectric layer interposed there in between. In one embodiment, said first electrically-conductive layer comprises a conductive film. Said second electrically-conductive layer may also comprise a conductive film. Indeed, it may comprise a monomolecular metal film. The dielectric layer may comprise periodic mesoporous organosilica or other suitable dielectric material.

Will this really be the end of the battery?

Keep in mind that EEStor's patent had written: "Technologies for replacement of electrochemical batteries". Yesterday, an independent analysis was completed and we received the news:

> *"Edward D. Golla, Ph.D. and Laboratory Director for Texas Research International, has officially certified that EEStor, Inc. test equipment, procedures, and techniques are capable of providing the EEStor, Inc. required testing accuracy for EEStor's chemicals and powder production processing analyses."*

It appears that EEStor has achieved success on one of its most critical milestones - the certification of the completion of the powder crystallization of the constituents utilized in producing its CMBT powders. The specifics: The percentage of the constituents crystallized in the CMBT powders ranged from 99.57% to 100.00%, with the average being 99.92%. The report stated that, "This level of crystallization provides the path for the possibility that EEStor can provide the published energy storage for present products and major advancements in energy storage for future products."

"These key certified production milestones of particle crystallization, size, purity, and polarization are expected to assist EEStor in providing not only present and future energy store requirements, but also production consistency."

Zenn motor company's strategic energy storage partner, eestor, inc. Confirms progress towards commercialization of its technology, toronto, ontario – july 30, 2008

ZENN Motor Company Inc.'s (TSXV: ZNN; the "Company") goal to become the global leader in the development of zero emission vehicles, including highway capable electric vehicles, has moved a step closer as its strategic partner, EEStor, Inc., announced Certification of Additional Key Production Milestones and Enhancement of Chemical Purity in a press release issued on **July 29, 2008.**

"Yesterday's announcement bodes well for EEStor's completion of its third party verified permittivity milestone and is a very strong affirmation of our investment in and the rapid progress of our business plan," stated Ian Clifford, Chief Executive Officer. "Energy Storage has always been the Achilles heel to the mass adoption of electric vehicles and EEStor technology is the "better battery" that the world has been waiting for. Every technology and industry reliant on electrical energy storage will potentially benefit from this."

Repealing the Overpowering Trend - via Ultra

capacitors and a Chorus Motor

Conventional vehicles are designed with powerful internal combustion engines (ICE's) that are intentionally over-engineered and overpowered to produce large amounts of torque at critical yet infrequent times, usually 10-15 seconds at a time.

An example would be starting from a complete stop to merge onto a highway. However, a Series Hybrid, outfitted with the new Chorus Meshcon^{tm} motor/drive and using ultra capacitors in the design allows for reversing this overpowering trend.

By incorporating Chorus Meshcon, which can provide 10x the 'burst' torque when powered by an ultra capacitor, a Series Hybrid can compete in both "price" and "performance" in the mass market due to the unusual features of this new and unique motor. With these two technologies, a Series Hybrid vehicle can be designed around its 'average' power requirement and not its 'burst' power requirement, while still maintaining performance and achieving a much higher gas mileage.

In a Series Hybrid design, the internal combustion engine does not drive a heavy mechanical transmission, but instead directly produces electrical power with an attached alternator, and the resulting electrical output powers the electric motors and/or charges the battery and capacitor bank. The 'burst' torque is provided not from the ICE, but from a special and new type of electric motor, the Chorus(R) motor that is making its way from the aerospace industry to the automotive industry, and will allow Series Hybrids to realize significant gas mileage gains.

In the near future, our starting system may also incorporate a high density, stable and reliable tantalum-equivalent capacitor to start our permanent magnetic motor. Solar power and the permanent magnetic motor may also store their excess energy in the capacitors.

SAN FRANCISCO (Business 2.0 Magazine) -- **The Disruptor:** EEStor

The Innovation: A ceramic power source for electric cars that could blow away the combustion engine

The Disrupted: Oil companies and carmakers that don't climb aboard

Forget hybrids and hydrogen-powered vehicles. EEStor, a stealth company in Cedar Park, Texas, is working on an "energy storage" device that could finally give the internal combustion engine a run for its money, and begin saving us from our oil addiction. "To call it a battery discredits it", says Ian Clifford, the CEO of Toronto-based electric car company Feel Good Cars, which plans to incorporate EEStor's technology in its vehicles by 2008.

EEStor's device is not technically a battery because no chemicals are involved. In fact, it contains no hazardous materials whatsoever. Yet it acts like a battery in that it stores electricity. If it functions as it's supposed to, it will charge up in five minutes and provide enough energy to drive 500 miles on about $9 worth of electricity. At today's gas prices, covering that distance can cost $60 or more. The EEStor device would power a car for the equivalent of about 45 cents a gallon.

And we mean power a car. "A four-passenger sedan will drive like a Ferrari", Clifford predicts. In contrast, his first electric car, the Zenn, which debuted in August and is powered by a more conventional battery, takes hours to charge and can't go much faster than a moped.

The cost of the engine itself depends on how much energy it can store; an EEStor-powered engine with a range roughly equivalent to that of a gasoline-powered car would cost about $5,200. That's a slight premium over the cost of the gas engine and the other parts the device would replace -- the gas tank, exhaust system, and drive-train, but overall, getting rid of the need to buy gas should more than make up for the extra cost of an EEStor-powered car.

EEStor is tight-lipped about its device and how it manages to pack such a punch. According to a patent issued in April, the device is made of a ceramic powder coated with aluminum oxide and glass. A bank of these ceramic batteries could be used at "electrical energy stations" where people on the road could charge up.

EEStor is backed by VC firm Kleiner Perkins Caufield & Byers, and the company's founders are engineers Richard Weir and Carl Nelson. CEO Weir, a former IBM employee, won't comment, but his son, Tom, an EEStor VP, acknowledges, "That is pretty much why we are here today, to compete with the internal combustion engine." He also hints that his engine technology is not just for the small passenger vehicles that Clifford is aiming at, but could easily replace the 300-hp brutes in today's SUVs. That would make it appealing to automakers like GM and Ford, who are seeing sales of their gas-guzzling SUVs and pickup trucks begin to tank because of exorbitant fuel prices.

EEStor is a company based in Cedar Park, Texas, United States that has developed a superior type of capacitor for electricity storage, which EEStor calls 'electrical energy storage units' (EESU). The EEStor EESU is an integral part of the much-anticipated ZENNergy drive-train. Its CEO and president is Richard Weir, who is also a co-inventor named on their principal technology patent.

According to its patent, these units will use high-purity barium titanate coated with aluminum oxide and glass to achieve a level of capacitance claimed to be much higher than what is currently available in the market. The claimed energy density of the prototype is 1 MJ/kg; existing commercial super capacitors typically have an energy density around 0.02 MJ/kg, while lithium ion batteries are around 0.54–0.72 MJ/kg.[3] For comparison, gasoline has an energy density around 45 MJ/kg. However, the combustion process is unable to extract the majority of this energy and overall energy delivered to the wheel is typically less than 20% of this value; a value of 9 MJ/kg is more indicative. EEStor has stated that production examples should be about 2.5 MJ/kg, better than 1/4th the usable energy density of gasoline. If true, this represents an enormous leap in energy storage technology, which is the reason the company generates so much excitement and skepticism at the same time.

Based on these claims, a full charge should give the capacitor sufficient energy to drive a small car 300 miles (480 km). Although the technology should allow very fast charging (approximately 5 minutes), standard household wiring is not capable of delivering the power required for this, and therefore, charging times this rapid would probably require special-purpose, high-capacity dispensing stations. Overnight charging at home should still be practical, as is using a second EESU for the home which could be charged overnight using cheap, off-peak electricity to then charge the EEStor unit in the car in 5-10 minutes on demand - and deliver cheap electric power to the house as well, making expensive peak power plants obsolete. Also, according to Ian Clifford, a normal household outlet with 110 volt supply can fully charge the EEStor powered CityZENN in 4 hours for a 250-mile (400 km) range and a normal household outlet with 220 volt supply can fully charge the EEStor powered City ZENN in 2 hours, and obtain a similar range.

Chapter 3

Fuel Cells

The problem with fuel cells is that they still burn fuel and are limited by distance. The fuel is highly explosive and under all kinds of pressure. Unless they come up with an incredibly efficient fuel cell, we do not believe that this is the proper course of action. Besides, how sensitive are they to vibration? Vibration may also be a serious issue.

Fuel Cells, if cheap, clean, and carbon-free, may become a maxim of this millennium and an argument in favor of fuel cells. These devices may be used to produce electricity in homes and cars using oxygen from air and hydrogen from natural gas. The resulting waste product is harmless water vapor.

At ORNL, Tim Armstrong of the Metals and Ceramics (M&C) Division is leading the effort to develop solid-oxide fuel cells and components using advanced materials. This type of fuel cell is flexible in the fuels it can use. For example, it can use natural gas in a process to convert chemical energy to electrical energy. How is the system made and how does it work?

Hydrogen from natural gas is passed over an anode (negative electrode) made of nickel and yttria-stabilized zirconia (YSZ). The hydrogen atoms break apart into positively charged ions and electrons. The electrons travel through an external circuit to a cathode (positive electrode) made of the rare-earth oxide lanthanum strontium manganate ($LaSrMnO_3$).

Oxygen from the air (or carbon monoxide) is collected at the cathode where the gas accepts electrons to form negatively charged oxygen ions, which are passed through a YSZ electrolyte separating the anode and cathode. The electrolyte is heated to a temperature of up to 1000°C by an electric heater to start the electrochemical reaction. On arrival at the anode, each oxygen ion is discharged by reacting with two hydrogen ions to form water. The heat from the electrochemical reaction maintains the cell temperature, allowing the electric heater to be turned off.

Iron aluminide alloys developed in the M&C Division by C. T. Liu, Claudette McKamey, and Vinod Sikka are candidates for solid-oxide fuel cell containment vessels. ORNL Fossil Fuel Energy Program Manager Rod Judkins and Sikka worked with Siemens Westinghouse to confirm the efficacy of iron aluminide in the fuel cell application. Compared with the various types of stainless steel currently used for containment, the iron aluminide alloys are stronger and more resistant to the simultaneous oxidizing and reducing conditions to which containment vessels are exposed. Thus, iron aluminide containments are expected to be more reliable and to last much longer. Mike Santella of the M&C Division is working with industrial partners to develop the technology for fabricating iron aluminide containment vessels.

"This solid-oxide fuel cell can also provide high-quality waste heat that can be used to warm the home or provide refrigeration and air conditioning," Armstrong says. "Its only emissions are steam, trace amounts of nitrogen oxides and sulfur oxides, and a small amount of carbon dioxide."

For the purpose of powering many homes simultaneously while eliminating carbon dioxide emissions, the M&C group has designed a power plant using a solid-oxide fuel cell and gas turbine incorporating ORNL's novel activated-carbon and membrane-separation technologies and heat-exchange technologies, such as the carbon foam that rapidly transports heat. The most efficient gas turbines available today produce electricity using 60% of the energy in the gas. Siemens Westinghouse has designed a solid-oxide fuel cell that is 60% efficient and a hybrid fuel cell micro-turbine plant in which waste heat from the fuel cell is used to drive the micro-turbine. This combined-cycle power plant is 70% efficient.

"Based on the results of our computer model, the ORNL design for a similar combined-cycle power plant is 80% efficient," Armstrong says. "The reason is that we combine our efficient heat-exchange technologies with membrane technologies for separating hydrogen and carbon monoxide from natural gas for use in the fuel cell." In addition, the power plant allows carbon sequestration because ORNL's carbon-fiber composite molecular sieve technology can capture the carbon dioxide leaving the fuel cell as a waste product. This gas can then be collected and used for enhanced oil recovery or sequestered in geological formations.

Cars powered by electricity from hydrogen fuel cells are being designed because they will eliminate discharges of carbon dioxide, nitrogen oxides, and particulate emissions. Such a "zero emissions" vehicle is a goal of the Partnership for a New Generation of Vehicles, which involves the automobile industry and the Department of Energy.

A proton exchange membrane (PEM) fuel cell is the technology of choice in the automobile industry for future electric cars because of its low-temperature operation and rapid startup. PEM fuel cells have been plagued with problems, but recent developments at ORNL may make this technology more feasible and affordable.

The problem with using today's PEM fuel cells to power cars is that their bipolar plates (positive and negative electrodes), which are made of machined graphite, are too heavy, too brittle, and too costly for use in automobiles. ORNL's solution is to make bipolar plates from a carbon-fiber composite, which is lighter, tougher, and cheaper than machined graphite.

ORNL's carbon-composite bipolar plate may be used for fuel cells in electric cars.

Ted Besmann, James Klett, Tim Burchell, and John J. Henry, Jr., all of the M&C Division, have developed a method for making composite plates that includes chemical vapor infiltration. Basically, carbon fibers are molded to make an electrode, and methane is flowed over the plate at high temperatures to deposit carbon that seals its surface pores. Because a fuel cell is a stack of bipolar plates with electrolytes in between, the porous plate surfaces must be sealed to prevent leakage of hydrogen and oxygen from one cell to another—a showstopper for fuel cells.

ORNL researchers have shown that carbon-fiber composite plates not only can be made to perform as well as graphite plates, but are also half the mass, as little as one-fifth the cost, are more conductive and corrosion resistant, and are easier to manufacture.

Thanks to ORNL's progress in this area, the fuel-cell car may be just around the bend.

Chapter 4

Batteries

Silicon Energy Storage Battery

Synopsis of the Operating Mechanism:

The electrolyte of silicon energy adopts silicate composite; making its performance surpass sulfuric acid in all aspects. It eliminates acid fog during manufacturing and deployment. The silicon energy storage battery charges with an internal chemosynthesis method, which is different from the standard lead-acid storage battery.

The following are the operating mechanisms:

1. During the chemosynthesis charging and battery repetitive charging process, the electrolyte added with silicate composite transforms from a solid state to a liquid state. This improves the charging acceptability and lowers the temperature.
2. The above transformation phenomenon provides a better physical condition to overcome the electrolyte stratification. It enables the battery to have a bigger capacity and stronger electric current during charging/discharging.
3. After charging the battery, the silicate electrolyte will not have the silicon split effect. It overcomes the AGM diaphragm and the "islet"

effect of the electrode plate. It solves the problem of harm from thermal runaway effect.

The following are the properties of the silicon energy environmentally friendly storage battery:

- **High capacity**. It reaches 1.75 times of what international standards require.
- **Strong charging acceptance capability**. It reaches 2.68 times of what international standards require.
- **Strong electric current charges**. It reaches 0.8° Celsius to 1.0° Celsius.
- **Strong electric current discharge.** 30° Celsius can discharge in 8 seconds without damaging the battery.
- **High speed charging.** It can be 95% charged in 20 minutes and finish charging within one hour.
- **Low self-discharging**. After completely charging, and if stored at normal temperatures, it can still be used within one year.
- **High capacity in low temperatures.** Its capacity can be maintained at 70% to 75% at an environmental temperature of 10° Celsius.
- **Wide temperature range**. It can be used in environmental temperatures from -40° to +50° Celsius.
- **Charging is acid fog-free.** Electrolyte is pollution-free.
- **Long-lasting life**. As for the GM type, under normal utilizing conditions, its floating charging can run for 10 years.

The genuine maintenance-free battery.

A silicon battery carries electric charge when it comes out of the production line. It can be installed for use right after it is available, and there is no need for any maintenance. It is a genuine maintenance-free battery. The battery's valve control is hermetically sealed and is maintenance-free.

Outstanding property/price ratio.

A silicon energy battery's strong electric current instant discharging capability is 3 - 5 times that of a lead-acid battery. The mass power is 30% more than a lead-acid storage battery. The sale price of a silicon energy storage battery is more than 40% less than the imported maintenance-free lead-acid battery.

Sensitive to low-speed, smaller electric currents.

The internal resistance of the silicate battery is very low and the active material is also sensitive to the tiny electric currents.

Small internal resistance, charge and discharge with strong electric current.

The internal resistance of the silicate battery is very low. It's only 1/10 to 1/18 that of a lead-acid battery.

Can be used in a wide temperature range.

A silicon energy battery can function normally in wide temperature ranges from -40° Celsius to 50° Celsius. When the temperature reaches 70° Celsius, it can still provide electric energy.

Full discharge functionality.

The active material for electric storage in a silicate battery has a strong, active anti-aging property. It has full discharging functionality and there is no minimum discharge voltage limit. It can be charged at any time and there is no need to discharge completely before recharging. This advantage can save electric energy, and more importantly, it brings more convenience to the users of regenerative energy.

Excellent power density and energy density.

A silicon energy battery's power generation and energy capacity is higher than a lead-acid battery.

Environmental protection of silicon-energy battery.

A silicon-energy battery abandons electrolyte sulfuric acid and provides a good, clean environment for battery production.

National and industrial certificates of inspection and permits of network entrance received by silicon energy environmentally friendly battery.

To summarize, compared with a lead-acid battery, the silicon energy storage battery has obvious advantages in environmental protection, property and cost. With the gradual understanding of the silicon energy battery from the designers and manufacturers of wind power, electric generator systems and solar generator systems will come the potential for an upsurge in the use of the silicon energy battery. The silicon energy storage battery will certainly become the new favorite of regenerative energy sources.

Composite Foam Grid Batteries.

Firefly is using composite foam grids instead of copper lead plate acid. It doesn't fall apart and shed the way lead acid does. When you buy a battery, it is top-heavy. It is very easy to tip over. The bottom one-third of the battery is a void space where the waste silicates accumulate, which is why they are easily recycled. The new battery with foam grids will not have this problem. To date, this is the best current technology available for our magnetic motor.

In the advanced battery architectures that Firefly has perfected, the Microcell™ composite foam "grids" are impregnated with a slurry of lead oxides which are then formed up to the sponge lead and lead dioxide in the normal fashion. Because of the foam structure, the resultant negative and positive plates have enormous surface-area advantages over conventional lead-acid grid structures. This results in much-improved active material utilization levels (i.e., from the historical 20-50% up into the range of 70-90%), as well as enhanced fast-recharge capability and greater high-rate/low-temperature discharge times. This battery may very well drive the magnetic engine.

Lithium-Ion Hybrid Batteries.

We have a space and a weight issue. How many lithium-ion batteries would we need to take the place of a composite foam grid battery?

What is the life span between the composite foam grid battery and the lithium-ion hybrid batteries? There is no sense having a thousand of these batteries dispersed all around the frame of the car.

A Laptop in Every Garage.

The emergence of small, lightweight long-running lithium-ion batteries has helped create a market for notebook computers, cell phones, and other portable devices from the iPod to the BlackBerry. Now, efforts to scale that technology for use in car batteries could do for the automotive industry what it did for computer and phone companies. The benefit for consumers could be revolutionary; hybrid or pure electric cars with great efficiency, acceleration and range—at or below the price of today's conventional cars.

On Dec. 16, 2005, Toyota announced that they would accelerate development of lithium batteries for use in their hybrids. That's a clear sign of how important battery power may become for the auto industry. For Toyota to achieve its goals of selling one million hybrids globally per year and offering hybrid versions of all their popular vehicles, they'll need to reduce the incremental cost of their hybrids. Earlier this year, Katsuaki Watanabe, Toyota's president, articulated that goal as reducing "the cost difference of hybrids to one-half the current levels." According to Dave Hermance, executive engineer of environmental engineering at Toyota, batteries are the single most expensive element in the hybrid system. "To make a big leap forward with hybrids," he said, "we'll need a new battery technology."

The current crop of hybrid cars, like the Toyota Prius and the Honda Civic hybrid, use a combination of gasoline and nickel metal hydride (NiMH) batteries. NiMH is a major step up from the lead-acid battery that most drivers recognize under their hoods. Battery formulas with nickel, while more powerful than lead acid, have not provided the breakthrough and the long-term cost benefits that battery engineers had hoped. The power of nickel batteries comes from the raw material itself, which is getting more expensive due to increased demand, while the benefits of lithium batteries are derived from the mass processing, which can scale to the high volumes required for the rapidly growing hybrid market without a corresponding jump in price.

The emergence and adoption of advanced lithium batteries could shake up the auto industry. To better understand the changes ahead, several auto companies, including Subaru, Nissan, and Mitsubishi, have produced concept cars that use lithium batteries. Toyota became the first to use lithium-ion batteries in a production vehicle when it placed a four-cell lithium pack in its Vitz CVT4, a Scion-resembling, small, boxy vehicle only available in Japan. The Vitz's battery pack powers the lights, heater, air conditioner, and radio while the car is stopped, allowing the gasoline engine to shut off.

By far, the boldest vision of the lithium future is the Volvo 3CC concept car. Unveiled in 2004, the all-electric vehicle relies exclusively on 3,000 lithium-ion cells, each one approximately the size of a common "AA" battery, to provide the equivalent of 105 hp with absolutely no emissions. "We were forced by the current marketplace to use the small format batteries. If we could have done it with less than 3,000 cells, it would have made my life a lot easier," said Ichiro Sugioka, science officer at Volvo's Monitoring and Concept Center in Camarillo, California.

Sugioka was unable to use larger lithium-ion batteries more suitable for cars because they are uncommon in today's market. Battery companies trying to create larger lithium batteries, which use cobalt in their formula, quickly encounter a problem known as thermal runaway, which means the batteries can easily catch fire or explode. Not a good idea for cars. However, two enterprising companies, Valence Technologies and A123 Systems, have had better results by replacing cobalt with other metals, such as phosphate. Here's how Valence describes its approach: "Whereas traditional lithium-ion technology utilizes cobalt-oxide cathode material, Saphion® technology incorporates a phosphate based cathode material."

The result is inherently less powerful; however, these two companies' proprietary technology is working like magic to produce energy storage and charging power similar to cobalt, while avoiding the safety and longevity issue. Yet, the technology still has a way to go—perhaps three to five years—before car-sized lithium batteries are powerful enough, cheap enough, and most importantly, reliable enough to go into mass production vehicles. Pinpointing an exact date when the first carmaker takes the risk on the new batteries is tough. "The crystal ball on this is real cloudy, but we'll get there. Lithium ion has too many advantages," said John German, manager of Environmental and Energy Analyses for American Honda.

Business Chemistry

Dozens of smaller companies are racing toward a lithium payday. Some say that their lithium batteries will solve all our transportation problems, and make hydrogen fuel cells immediately irrelevant. The major car companies are not taking the wild claims very seriously, but some battery companies have attracted wider attention. In November, A123 Systems, a private company in Watertown, Mass., unveiled a lithium battery that earned them $32 million in funding from big name investors, such as Sequoia Capital, Motorola, and the Massachusetts Institute of Technology. The company is working with the Department of Energy on the development of lithium-ion batteries for hybrid vehicles. Lithium iron phosphate batteries from Valence Technology, a publicly traded company based in Austin, Tex., are being used in Segway electric scooters and a much-publicized, converted plug-in Prius.

Toyota has spent millions of dollars to make sure that consumers know that current hybrids don't need to be plugged in. However, many industry insiders predict that a hybrid with the option to plug in not only represents the next generation of hybrid vehicles, but a major shift in automotive technology. Using a standard home electric plug and outlet, a driver could recharge a car overnight during off-peak hours, and wake up with a vehicle ready to travel 50 or 100 miles without using a drop of gasoline. This cannot occur without more robust batteries (among other things), and the entire industry is looking at lithium batteries as the solution. EDrive Systems, a joint venture between EnergyCS Inc. and Clean-Tech LLC, has created a number of plug-in hybrid prototypes in its effort to commercialize the new technology. EDrive utilized Valence's lithium batteries.

The right chemical formula does not necessarily produce a winning business formula. In the first quarter of the 2006 fiscal year, Valence pulled in $3.4 million in revenue, a 20% increase from the previous year. However, the company, which was founded in 1989, has never recorded a profit. Valence, A123, and other smaller battery makers could greatly benefit from a partnership with a major car company. Dean Bogues, president of worldwide sales, marketing, and product development for Valence, said that commercialization of their lithium car batteries is not likely to occur overnight via a brand name car company, but will "unfold over time" through partnerships and distribution in commercial fleets. Saft, a battery maker based in France, has entered into such a partnership with DaimlerChrysler by deploying its lithium batteries for use in the car company's plug-in hybrid Sprinter van, currently being tested but with no clear plans for production.

Bogues added, "If there was interest from a major OEM [original equipment manufacturer], we wouldn't turn it down." He explained that it commonly takes 10 years to commercialize new battery chemistry, and Valence is currently about three years into that product development cycle.

Small battery companies generally lack experience with auto manufacturers, who can be very demanding from their suppliers. The G.M.'s, DaimlerChrysler's, Honda's, Ford's, and Toyota's of the world want a battery company to deliver a finished integrated system, not just the component cells, electronics, controls, etc. That problem trickles down to another set of players, like Altair Nanotechnologies, who serve the small battery makers with raw materials.

Progress from Within

Toyota, with big hybrid dreams, is not waiting for the small players to deliver. The Japanese maker of the Prius and Highlander hybrid recently spent $740 million to establish 9% ownership stake in Fuji Heavy Industries, which makes advanced hybrid batteries (and Subaru vehicles). In October, Toyota also increased their equity in Panasonic EV Energy from 40 to 60%. PEVE, the world's leading supplier of nickel-metal hydride, will act as the key player in Toyota's recently announced venture to develop a lithium-ion battery for hybrid cars.

Johnson Controls is also making a play in the hybrid battery market. In 2002, Johnson Controls acquired Varta, a major European automotive battery manufacturer. In September 2005, they launched a $4 million lab to accelerate work on lithium batteries, and in October, signed a memorandum of understanding to work with Saft. Johnson Controls could emerge as the only major U.S.-based hybrid battery maker. The United States is the world's largest hybrid car market, and all the car companies who manufacture hybrids in the U.S.—Toyota began production of the Camry hybrid in 2006—would benefit from having a domestic supplier. If Johnson Controls develops lithium expertise, they will be even better positioned.

Lithium and Gold

The average driver has little or no knowledge of battery chemistry. Small battery companies and large auto companies know exactly what it means: a critical step towards transforming hybrid cars from a niche curiosity to a major revolution in automotive technology. Lithium batteries could be the key. Not too long ago, we woke up one day and found ourselves suddenly surrounded by cell phones and laptops. We're getting closer and closer to the day—probably just beyond the end of this decade—when one or more battery companies hit the right formula and lithium batteries for cars are produced in the millions. Engineers in those companies are like modern day alchemists, trying to turn base materials into gold. If they succeed, they could indeed be very rich.

The historical alchemists had deeper aspirations. They sought personal spiritual transformation. Modern day battery alchemists also aspire to loftier goals: a world of personal mobility without auto pollution, global warming, or oil wars.

Chapter 5

Solar Power

The average person commutes to work and puts in 8-10 hours. With the car parked under the sun, the battery has all that time to be maintained. They now have the ability to paint the photovoltaic cells right into the surface of the vehicle. There are epoxy coatings that can be spray painted onto the car that act as an insulator. The solar cells can be spray painted onto the epoxy surface as stripes and connected with small copper jump wires, forming a grid with the proper output to maintain the batteries.

A solar vehicle is an electric vehicle powered by solar energy obtained from solar panels on the surface of the vehicle. Photovoltaic (PV) cells convert the sun's energy directly into electrical energy. Solar vehicles are not practical day-to-day transportation devices at this time, but are primarily demonstration vehicles and engineering exercises, often sponsored by government agencies.

Solar cars combine technology typically used in the aerospace, bicycle, alternative energy, and automotive industries. The design of a solar vehicle is usually severely limited by the energy input into the car (batteries and power from the sun). Virtually all solar cars ever built have been for the purpose of solar car races (with notable exceptions).

Like many race cars, the driver's cockpit usually contains room for only one person, although a few cars do have room for a second passenger. They contain some of the features available to drivers of traditional vehicles such as brakes, accelerator, turn signals, rearview mirrors (or camera), ventilation, and sometimes cruise control. A radio for communication with their support crews is almost always included.

Solar cars are often fitted with gauges, as seen in conventional cars. Aside from keeping the car on the road, the driver's main priority is to keep an eye on these gauges to spot possible problems. Cars without gauges available for the driver will almost always feature wireless telemetry. Wireless telemetry allows the driver's team to monitor the car's energy consumption, solar energy capture, and other parameters and allow the driver to freely concentrate on driving.

Tesla Motors is offering a modestly-sized and priced solar panel from SolarCity. It can be installed on the car roof in an out-of-the-way location because of its small size, or set up as a carport, which will generate about 50 miles per day of renewable electricity.

The electrical system is the most important part of the car's design as it controls all of the power that enters and exits the system. The battery pack plays the same role in a solar car as a fuel tank plays in a normal car in storing power for future use. Solar cars use a range of batteries including lead-acid batteries, nickel-metal hydride batteries (NiMH), nickel-cadmium batteries (NiCd), lithium-ion batteries and lithium-polymer batteries.

Many solar race cars have complex data acquisition systems that monitor the entire electrical system, while even the most basic cars have systems that provide information on battery voltage and current to the driver.

The mechanical systems of a solar car are designed to keep friction and weight to a minimum while maintaining strength. Designers normally use titanium and composites to ensure a good strength-to-weight ratio.

Solar cars usually have three wheels, but some have four. Three-wheelers usually have two front wheels and one rear wheel: the front wheels steer and the rear wheel follows. Four-wheel vehicles are set up like normal cars or similarly to three-wheeled vehicles with the two rear wheels close together.

The solar array consists of hundreds of photovoltaic solar cells converting sunlight into electricity. Cars can use a variety of solar cell technologies, most often polycrystalline silicon, monocrystalline silicon, or gallium arsenide.

The power produced by the solar array depends on the weather conditions, the position of the sun, and the capacity of the array. At noon on a bright day, a good array can produce over 2 kilowatts (2.6 hp).

Some cars have employed free standing or integrated sails to harness wind energy.

Solar cars from the University of Michigan and University of Minnesota heading west toward the finish line in the 2005 North American Solar Challenge.

The two most notable solar car races are the World Solar Challenge and the North American Solar Challenge, overland road rally-style competitions contested by a variety of university and corporate teams.

The World Solar Challenge features a field of competitors from around the world who race to cross the Australian continent, over a distance of 3000 km. The increasingly high speeds of the 2005 race participants led to the rules being changed for future solar cars starting in the 2007 race.

The North American Solar Challenge, previously known as the 'American Solar Challenge' and 'Sunrayce USA', features mostly collegiate teams racing in timed intervals in the United States and Canada. The last North American Solar Challenge ran from June 13-22, 2008, from Dallas, Texas to Calgary, Alberta.

In 2006, the Venturi Astrolab was hailed as the world's first commercial electro-solar hybrid car due to be released in January 2008. In May 2007, a partnership of Canadian companies led by Hymotion, altered a Toyota Prius to use solar cells to generate up to 240 watts of electrical power in full sunshine. This is reported as permitting up to 15 km of extra distance on a sunny summer day while using only the electric motors.

Battery electric vehicles fitted with solar cells would extend their range and allow for recharging while parked anywhere in the sun. However, with present and near-term engineering considerations, it seems that the more likely place for solar cells will generally be on the roofs of buildings, where they are always exposed to the sky and weight is largely irrelevant, rather than on vehicle roofs, where size is limited. Energy from rooftop panels can be stored in batteries for future use.

CHAPTER 6

THE HISTORY OF MAGNETIC MOTORS

Roy Meyers' Absorber -- "Free Energy"

by Robert A. Nelson

As planet Earth suffers under the relentless onslaught of ever increasing human consumption, we are becoming increasingly desperate for sources of energy with which to power our pretensions of civilization. There is not enough oil or coal available to sustain the present rate of growth, and the apparently inevitable result will be endless wars over dwindling resources. There are, however, numerous marvelous alternative energy technologies that have been suppressed by the powers that be.

We could easily be enjoying the benefits of clean, cheap (and even free) energy in various forms, were it not for the insatiable greed of our corporate masters, who rule the world without regard for humanity or the environment.

There are dozens of stories that can be told of inventors who presented wonderful liberating technologies, only to be ruined for their efforts. Many people have heard of various types of so-called "free energy" or "over-unity" generators that tap what scientists call Zero Point Energy (and other names, such as the aether) without need for fuel and without producing any pollution. Detractors like to label such devices as impossible "perpetual motion machines", even though some of them are solid state electronic systems that have been built and shown to work. Readers who are interested in such possibilities need search no farther than the impeccable history of Thomas Henry Moray, which is widely published on the internet (e.g., http://www.rexresearch.com/moray.moray.htm).

If that is too much bother for you, then just continue reading here. This is the little-known story of **Roy Jerome Meyers**, who invented an "Absorber" that produced limitless electrical power from the terrestrial magnetic field.

Almost 100 years ago, Meyers served a 3-1/2 year sentence in the Arizona state prison at Florence. Before his incarceration, he had already invented an improved trolley wheel head that prevented it from jumping off the wire. While he was imprisoned, he was inspired to invent a device that would draw electricity from the atmosphere. He conferred with Superintendent Sims and Parole Clerk Sanders, and convinced them to give him the opportunity to develop his idea in a wooden shed on the grounds of the penitentiary. Within a few weeks, using easily available materials (chrome steel magnets and iron wire), he constructed the first crude working model, and used it to spark the gas engines of the prison's pump house. His second model developed 8 volts.

Miss Kate Barnard, who was State Commissioner of Charities and Corrections of Oklahoma, was a guest of Superintendent Sims at the time, and she saw the machine in operation. Miss

Barnard was so impressed by it and by Meyers' essential integrity (despite the lapse that had gotten him imprisoned) that she told the story of Roy Meyers when later she appeared before the Arizona legislature to address them concerning prison reform. The legislature and Governor Hunt were convinced to grant Meyers an unsupervised leave of absence for 30 days to travel to Washington, D.C. in order to apply for a patent.

Meyers gave this account of his trip:

"When I arrived in Washington and laid my plans before the patent office experts, they merely smiled and told me that I would have to build a model and demonstrate my claims --- that it seemed strange that I, unknown as I am in the electrical world, should have accomplished the things for which Edison, Tesla and other experts have been striving for years.

"They could not grasp the meaning of my drawings nor the explanation I tried to make to them. There was little time to spare, as I had only 20 days left of my leave, but I set to work in a few days was able to take a crude model around to the patent office to make a demonstration.

"Arriving at the patent office, I telephoned to a friend who had been so kind as to introduce me and aid me in reaching the proper officials. The absorber was hoisted on two short poles and made to work. While they were as yet unable to understand the principles involved and hardly willing to believe their eyes, they were forced to admit that I had something new and different, and they told me that there would be no further objection; that I might file my application without further delay.

"I hope to construct my first large machine right here in Phoenix. I feel grateful; to Governor Hunt and others for what they have done for me and to the help they have given in securing protection I might not otherwise have had, and I am desirous of demonstrating this gratitude. I am going back to Florence today to resume the serving of my sentence, which will expire in 10 months. Then, here in Phoenix, I will begin the work of making my machines." (*Technology World Magazine,* 1912)

Meyers planned to build a 200 foot tower upon which to place an "Absorber" (as he called it) with sufficient power to light the city of Phoenix. According to him, it comprised "a series of magnetized steel plates set in a circle and this mechanism attracts the electricity from the atmosphere. This is carried by wires to a transformer in the engine house below and thence is applied to produce either power or light after the usual manner.

"The flow of electricity is constant. When it emerges into the transformer, it is in the form of a direct current. It will absorb the electricity day and night and will work whenever the wireless will work. I can put up a plant to supply such a building as the Adams Hotel for about $1500, and one of the principal items of the expense is the cost of the towers, the wires, the magnetizing of one set of plates, which is part of the secret of the treatment which makes it respond to the accumulations of the atmosphere.

"For use in the case of an electrical storm, I have made what I call a modified form of circuit breaker, such as is commonly used as a lightning arrester on telegraph lines. In case of a storm, the accumulator would suddenly become overcharged, possibly, and as the electricity would not of itself flow back into the air, the result might be disastrous. So I send it down into the ground, whenever the voltage rises above a certain amount."

And the story ends there. It is not known what became of Roy J. Meyers. No "Absorber" was built to power Phoenix, and it appears that a US Patent was not granted. This writer first learned of Meyers' Absorber in 1986, and repeatedly searched for it in the US Patent Gazette, to no avail. It was not until 2006 that Meyers' British patent for "Improvements in and Relating to Apparatus for Producing Electricity" was located after several periodic advanced searches of the European Patent Office online (http://ep.espacenet.com). The patent has long since expired and entered the public domain. It is reprinted below for your edification.

It should be obvious from this brief but indisputable account, and the description given in the patent, that "free energy" is a reality. Anyone who thinks otherwise is merely ignorant or in denial.

Roy J. Meyers & the "Absorber"

References

(1) Charlton Lawrence Edholm: "Picks Power from the Air"; *Technology World Magazine*, p. 279-281 (1912) *Electronic World* (November 1912)

(2) Roy Jerome Meyers: British Patent # 1098 (January 14, 1913); Improvements in and Relating to Apparatus for Producing Electricity

Magnetic Motor

The magnetic motor will have no wear surfaces. We have a dual thrust bearing that maintains armature position. The armature itself is supported on air bearings. There are no brushes or armature to wear out. The magnetic motor should be the least of our worries for preventive maintenance.

H.R. Johnson patent No. 4,151,431 on April 24, 1979 developed a permanent magnet motor. Mr. Johnson arranged a series of permanent magnets on the rotor and a corresponding series – with different spacing – on the stator. One simply has to move the stator into position and rotation of the rotor begins immediately.

The invention is directed to the method of utilizing the unpaired electrons that spin in Ferro magnetic fields and other materials as a source of producing power without any electron flow, as occurs in normal conductors, and to permanent magnetic motors for utilization of this method to produce a power source. In the practice of this invention, the unpaired electron spins occurring within permanent magnets are utilized to produce a motive power source solely through the super-conducting characteristics of a permanent magnet. The magnetic flux created by the magnets are controlled and concentrated to orient the magnetic forces generated in such a manner to do continuous work, such as the displacement of a rotor with respect to a stator. The timing and orientation of magnetic forces at the rotor and stator component produced by permanent magnets in order to produce a motor is accomplished with the proper geometrical relationship of these components.

Before you dismiss the idea of a magnetically run motor, a free energy source, consider the following: Engineers of Hitachi Corp. of California have stated that a motor run solely by magnets is feasible and logical, but the politics of the matter make it impossible for them to pursue the development of a magnet motor or any device that would compete with the energy cartels.

In a book titled *Keely and His Discoveries,* published in 1893 by Clara B. Moore, we find the following statements: A magnet that lifts a pound one day, if the load is gradually increased daily, that same magnet will lift double that amount with time. From whence comes this energy? Keely teaches that it comes from sympathetic association with one of the currents of the polar stream and that its energy increases as long as the sympathetic flow lasts, which is through eternity. Now consider some basic observations concerning magnets: 1). Two permanent magnets can either attract or repel, depending on the arrangement of the magnetic poles. 2). Two magnets repel further than they attract because of friction and inertia forces. 3). Most of our energy comes directly or indirectly from electromagnetic energy of the sun, e.g., photosynthesis and water cycles of ocean to water vapor to rain, or snow to ocean. 4). Magnetic energy travels between poles at the speed of light. 5). Permanent magnets on both sides of an iron shield are strongly attracted to the shield and only weakly attracted to each other at close proximity to the shield. 6). Permanent magnets are ferrous metals and are attractive only. Attraction is an inverse square force. 7). Magnetic energy can be shielded. 8). The sliding or perpendicular force of a keeper is much less than the force in the direction of the field to remove the keeper. 9). Most of the magnetic energy is concentrated at the poles of the magnet. 10). A permanent magnet loses little strength unless dropped or heated. Heating misaligns the magnetic elements within the magnet. 11). If a weight lifted by a permanent magnet is slowly increased, the lifting power of the magnet can be increased until all the magnetic domains in the magnet are aligned in the same direction, which becomes the limit. 12). Using magnets to repel tends to weaken them as it causes more misalignment of the domains. 13). A magnetic material placed between two magnets will always be attracted to the stronger magnet. Therefore, our ultimate motor becomes a permanent magnet motor of proper size with speed being controlled via the automobile's transmission. Additionally, and here is the biggest plus; permanent magnets keep their strength for a minimum of 95 years! As a result, we have a fuel–less automobile that would last us our lifetime. There is only one drawback to an automobile powered by a permanent magnet motor – if the vehicle gets involved in an accident, the shock of the crash could jar the magnets and cause them to lose power! This seems to be a small price to pay for an automobile that could run all day at 60 mph, without using fuel, and never requiring a recharge! The only question left to be answered is: 'Where do you buy one?' Or perhaps, 'When will we be able to buy one?' Currently, there are several companies offering interim solutions. Some offer electrically- powered designs, although this is strictly batteries, while others offer a hybrid combination of batteries and small gasoline engines. All of these so-called modern alternatives suffer from the same lack of accessories which we've become accustomed to. They do not, and cannot offer power steering, brakes, windows, air-conditioning, etc. Since they are small aerodynamically shaped packages holding only two people, their appeal is distinctly limited. When someone constructs an automobile run by a permanent magnet motor attached to the differential thus eliminating the transmission, the world will beat a path to his door – provided that the energy cartel doesn't find him first!

"We don't grant patents on perpetual motion machines," said the examiners at the U.S. Patent Office. "It won't work because it violates the law of conservation of energy," said one physicist after another. Since inventor Howard Johnson is not the sort of man to be intimidated by such seemingly authoritative pronouncements, he now owns U.S. Patent No. 4,151,431, which describes how it is possible to generate motive power, as in a motor, using only the energy contained in the atoms of permanent magnets. That's right. Johnson has discovered how to build motors that run without an input of electricity or any other kind of external energy!

The monumental nature of the invention is obvious, especially in a world facing an alarming, escalating energy shortage. Interestingly, inventor Johnson is not rushing to peddle his creation as the end-all solution to worldwide energy problems. He has more important priorities. First, there's the need to refine his laboratory prototypes into workable practical devices, particularly a 5,000-watt electric power generator already in the building. His second and perhaps more difficult major challenge; persuade a host of skeptics that his ideas are indeed practical.

Johnson, who has been coping with nonbelievers for decades, can be very persuasive in a face-to-face encounter because he cannot do more than merely theorize; he can demonstrate working models that unquestionably create motion using only permanent magnets. When this writer was urged by the editor of *Science & Mechanics* to make the thousand-mile pilgrimage to Blacksburg, Virginia to meet with the inventor, he went there as an "open-minded skeptic" and as a former research scientist determined not to be fooled. Within two days, this former skeptic had become a believer. Here's why.

Doing the Unthinkable.

Howard Johnson refuses to view the "laws" of science as somehow sacred, so doing the unthinkable and succeeding is second nature to him. If a particular law gets in the way, he sees no harm in going around it for a while to see if there's something on the other side. Johnson explains the persistent opposition he experiences from the established scientific community like this: "Physics is a measurement science and physicists are especially determined to protect the 'law' of conservation of energy. Thus, the physicists become game wardens who tell us what 'laws' we can't violate. In this case, they don't even know what the game is. But they are so scared that I and my associates are going to violate some of these laws that they have to get to the pass to head us off!"

The critics say Johnson offers a "free lunch" solution to energy problems, and that there can be no such thing. Johnson demurs, reminding repeatedly that he has never suggested that his invention provides something for nothing. He also points out that no one talks about a "free lunch" when discussing extraction of enormous amounts of atomic power by means of nuclear reactors and atom bombs. In his mind, it's basically the same thing.

Johnson is the first to admit that he doesn't actually know from where the power he has tapped derives. However, he postulates that the energy may be associated with spinning electrons, perhaps in the form of a "presently unnamed atomic particle." How do other physicists react to Johnson's suggestion that there may be an atomic particle thus far overlooked by nuclear physicists? Johnson states: "I guess it's fair to say that most of them are revolted." On the other hand, a few converted scientists, including some who are associated with large and prestigious research laboratories, are intrigued enough to suggest that there should be a hunt for the answer, be it a "particle" or some other as yet unsuspected characteristic of atomic structure.

This article is prefaced with the foregoing brief summary of the ongoing controversy so that in fairness to the inventor, we might all view his claims with open minds, even if it means temporarily setting aside cherished scientific concepts until more complete explanations are forthcoming. The main question to be answered here and now is this: Does Johnson's permanent magnet motor work?

Before providing the answer, we need to face up to another question that undoubtedly nags in the minds of many readers: Is Johnson a bona fide researcher, or merely a mad "garage mechanic" inventor? As the following brief summary suggests, the inventor's credentials appear to be impeccable. Following seven years of university training, Johnson worked on atomic energy projects at Oak Ridge, did magnetic research for Burroughs Company, and served as scientific consultant to Lukens Steel. He has participated in the development of medical electrical products, including injection devices. For the military, he invented a ceramic muffler that makes a portable motor generator silent at 50 feet. This has been in production for the past 18 years. His contributions to the motor industry include: a hysteresis brake, non-locking brake materials for anti-skid application, new methods of curing brake linings, and a method of dissolving asbestos fibers. He has also worked on silencers for small motors, a super charger, and has perfected a 92-pole no-brush generator to go in the wheel of Lincoln automobiles as a skid control; that last item reduced the cost to one-eighth of the cost of an earlier design by utilizing metal-filled plastics for the armature and field. In total, Johnson is connected with more than 30 patents in the fields of chemistry and physics.

Sticky Tape Scientist.

Despite his impressive credentials, this amiable and unpretentious inventor likes to characterize himself as a "sticky tape" scientist. He sees no virtue in wasting time building fancy, elaborate equipment when simpler assemblies serve just as well to test new ideas. The prototype devices shown in the photographs in this article were assembled with sticky tape and aluminum foil, the latter material being used mainly to keep individual, permanent magnets packaged together so that they do not fly apart.

The first item consists of more than a dozen foil-wrapped magnets assembled to form a broad arc. Each magnet is extended upward slightly at each end to form a low U-shape, useful for concentrating magnetic fields where they are needed. The overall curvature of the mass of magnets apparently has no particular significance except to show that the distance between these stator magnets and the moving vehicle is not critical. A transparent plastic sheet atop this magnet assembly supports a length of plastic model railroad track. The vehicle, basically a model railroad flatcar, supports a foil-wrapped pair of curved magnets, plus some sort of weight, in some cases, a rock. The weight is needed to keep the vehicle down on the track, against the powerful magnetic forces that would otherwise push it askew. That's all there is to the construction of this representation of a "linear motor."

The moment the inventor let go of the vehicle he carefully placed at one end of the track, it accelerated and literally zipped from one end to the other and flew off the track and onto the floor! Wow!

Powerful magnetic forces were at work as they placed the vehicle on the track. They gently eased the vehicle to the critical starting point, taking great care not to exert any kind of forward push, even inadvertently. They let go, and zip! It was on the floor again, at the other end of the track. Knowing that they would be asked if the track might've had a slant, They reversed the vehicle and started it from the opposite end of the track. It worked just as effectively in the reverse direction. In fact, the vehicle can even navigate a respectable incline. In light of these tests, and considering the remarkable speed of the vehicle, you can discount any notion that this was a simple "coasting" effect.

The second device has the U-shaped magnets standing on end in a rough circular arrangement oddly reminiscent of England's Stonehenge. This assembly is mounted on a transparent plastic sheet supported on a plywood panel that pivots underneath on a free turning wheel obtained from a skateboard. As instructed, they eased the 8-ounce focusing magnet into the ring of larger magnets, keeping it at least four inches away from the ring. The 40-pound magnet assembly immediately began to turn and accelerated to a very respectable rotating speed which it maintained for as long as the focusing magnet was held in the magnetic field. When the focusing magnet was reversed, the large assembly turned in the opposite direction.

Since this assembly is clearly a crude type of motor, there's no doubt that it is indeed possible to construct a motor powered solely by permanent magnets.

The third assembly, which resembles the bones of some prehistoric sea creature, consists of a tunnel constructed of rubber magnet material that can easily be bent to form rings. This was one of the demonstration models that Johnson took to the U.S. Patent Office during his appeal proceedings. Normally, the patent examiners spend only a few minutes with each patent applicant, but played with Johnson's devices for the better part of an hour. As the inventor was leaving, he overheard one sideline observer remark: "How would you like to follow that act?!"

It took Johnson about six years of legal hassling to finally obtain his patent, and he has been congratulated for his ultimate victory over patent office bureaucracy as well as for his innovative prowess. One indicator that he left the patent office more than a little shaken by the experience was the inclusion of diagrammatic material in the printed patent that does not belong there. If you were to look up the patent, pay no attention to the "ferrite" graph on the first page; it belongs in some other patent!

Of course, the tunnel device worked very well in the inventor's office during their visit, although Johnson observed that the rubber magnets are perhaps a thousand times weaker than the cobalt samarium magnets used for the other assemblies. There's just one big problem with the more powerful magnets: they cost too much. According to the inventor, the magnets used to construct the Stonehenge rotating model are collectively worth more than one thousand dollars. Fortunately, there's no need to depend solely on mass-production economies to bring the cost down to competitive levels. Johnson and U.S. Magnets and Alloy Co. are in the process of developing alternative, relatively low-cost magnetic materials that perform very well.

How do they work?

A curved "arcuate" armature magnet in three successive positions over a line of fixed stator magnets provides highly simplified insights into the theory of permanent magnet motive power generation. Johnson says curved magnets with sharp leading and trailing edges are important because they focus and concentrate the magnetic energy much more effectively than blunt-end magnets. These arcuate magnets are made slightly longer than the lengths of two stator magnets plus the intervening space, and in Johnson's setups, about 3-1/8" long.

The stator magnets all have their north faces upward and that they are resting on a highly magnetic permeability support plate that helps concentrate the force fields. The optimal gap between the end poles of the armature magnet and the stator magnets appears to be about 3/8".

As the armature north pole passes over a magnet, it is repelled by the stator north pole, and there's an attraction when the north pole is passing over a space between the stator magnets. Of course, the exact opposite is true with respect to the armature south pole. It is attracted when passing over a stator magnet, repelled when passing over a space.

The various magnetic forces that come into play are extremely complex. Solid lines represent attraction forces, dashed lines represent repulsion forces, and in each case, double lines indicate the more dominant forces.

The leading (N) pole of the armature is repelled by the north poles of the two adjacent magnets. Yet, at the indicated position of the armature magnet, these two repulsive forces (which obviously work against each other) are not identical; the stronger of the two forces (double dashed line) overpowers the other force and tends to move the armature to the left. This left movement is enhanced by the attraction force between the armature north pole and the stator south pole at the bottom of the space between the stator magnets.

Wait, there's more! Let's see what is happening simultaneously at the other end (S) of the armature magnet. The length of this magnet (about 3-1/8") is chosen in relation to the pairs of stators in magnets, plus the space between them so that once again, the attraction/repulsion forces work to move the armature magnet to the left. In this case, the armature pole (S) is attracted by the north surfaces of the adjacent stator magnets, but because of the critical armature dimensioning, is more strongly attracted by the magnet (double solid line) that tends to "pull" the armature to the left. It overpowers the lesser "drag" effect of the stator magnet to the right.

Additionally, in this case, there is the added advantage of repulsion force between the south pole of the armature and the south pole in the space between the stator magnets.

The importance of correct dimensioning of the armature magnet cannot be over-emphasized. If it's either too long or too short, it could achieve an undesirable equilibrium condition that would stall movement. The objective is to optimize all force conditions to develop the greatest possible off-balance condition, but always in the same direction as the armature magnet moves along the row of stator magnets. However, if the armature is rotated 180° and commenced at the opposite end of the track, it would behave in exactly the same manner, except that it would, in this example, move from left to right. Note also that once the armature is in motion, it has momentum that helps carry it into the sphere of influence of the next pair of magnets where it gets another push and pull, as well as additional momentum.

Complex Forces

Some very complex magnetic forces are obviously at play in this deceptively simple magnetic system, and at present, it is impossible to develop a mathematical model of what actually occurs. However, computer analysis of the system, conducted by Professor William Harrison and his associates at Virginia Polytechnic Institute (Blacksburg, VA.) provides vital feedback information that greatly assists in the effort to optimize these complex forces to achieve the most efficient operating design possible.

As Professor Harrison points out, in addition to the obvious interaction between the two poles of the armature magnet and the stator magnets, many other interactions are also in play. The stator magnets affect each other and the support plate. Magnet distances and their strengths vary despite best efforts of manufacturers to exercise quality controls. In the assembly of the working model, there are inevitable differences between horizontal and vertical air spaces. All of these interrelated factors must be optimized, which is why computer analysis in this refinement stage is vital. It's a kind of information feedback system. As changes are made in the physical design, fast dynamic measurements are made to see whether the expected results have actually been achieved. The new computer data is then used to develop new changes in the design of the experimental model, and so on.

The fact that very different magnetic conditions exist at the two ends of the armature is shown by the actual experimental data displayed in the table and associated graph. To obtain this information, the researchers first passed the probe of an instrument used to measure magnetic field strengths over the stator magnets and the intervening spaces. We shall call this the "zero"

level, although there is a very tiny gap between the probe and the tops of the stator magnets. These measurements effectively indicate what each pole of the armature magnet "sees" below as it passes over, i.e., the stator magnets.

Next, the probe is moved to a position just beneath one of the armature poles, at the top of the 3/8"-wide armature-to-stator air gap. Another set of magnetic flux measurements is made. The procedure is repeated with the probe positioned just beneath the other armature pole.

"Instinct" might suggest, and correctly so, that the flux measurements at the top and bottom of the air gap will differ. However, if "instinct" also suggests that these differences are relatively the same at the two armature pole positions, you would be very much in error!

To begin, study the actual flux density measurements. Note that in this particular experiment, the total magnetic flux amounted to 30,700 gauss (the unit of magnetic strength) when the probe was held at the "zero" level under the north pole of the magnet, and a total of 28,700 gauss when the probe was moved to the top of the 3/8"-wide air gap. The difference between these total measurements is 2,000 gauss.

Similar readings occurring at the air gap between the south pole of the armature and the stator magnets indicate a total flux at "zero" level of 33,725 gauss, and 24,700 gauss at the top of the air gap. This time, the difference is a much larger 9,025 gauss, or four and one half times greater than for the north pole! Clearly, the magnetic force conditions are far from identical at the two ends of the armature magnet.

The middle five pairs of figures from each table have been plotted in graphic form to make these differences more obvious. In the top "south pole" graph, the dashed line connects the "zero" level readings made over the stator magnets and over the intervening air spaces. Points along the solid line indicate comparable readings made with the probe just beneath the armature south pole. It is easy to see that there is an average attraction reduction of 43% between the armature and stator magnets created by the air gap. Equally true, but perhaps not so obvious, is the fact that there is an average repulsion increase of 36% when the south pole of the armature passes over the spaces between the stator magnets. The percentage increase only seems smaller because it applies to a much smaller "zero" level value.

The second graph shows that the changes are much less dramatic at the north pole of the armature. In this case, there's an average 11.7% decrease of attraction over the spaces, and a 2.4% increase of repulsion when the armature north pole passes over the stator magnets.

As you study the data, be sure to note that the columns are labeled differently. In the case of the north pole data, the stator magnet areas repulse the armature north pole while the spaces between the stator magnets attract. The conditions are exactly the opposite for the south pole of the armature magnet. When the south pole passes over a magnet, there is strong attraction, and when it passes over a space, there is repulsion.

The Ultimate Motor

A motor based on Johnson's findings would be of extremely simple design compared to conventional motors. As shown in the diagrams developed from Johnson's patent literature, the stator/base unit would contain a ring of spaced magnets backed by a highly magnetic permeability sleeve. Three arcuate armature magnets would be mounted in the armature which has a belt groove for power transmission. The armature is supported on ball bearings on a shaft that either screws or slides into the stator unit. Speed control and start/stop action would be achieved by the simple means of moving the armature toward and away from the stator section.

There is a noticeable pulsing action in the simple prototype units that may be undesirable in a practical motor. The movement can be made smoother, the inventor believes, by simply using two or more staggered armature magnets as shown in another drawing.

What's Ahead?

For inventor Howard Johnson and his permanent magnet power source, there's bound to be plenty of controversy, but also progress. A 5,000-watt electric generator powered by a permanent magnet motor is already on the way, and Johnson has firm licensing agreements with at least four companies at this writing.

Will we see permanent magnet motors in automobiles in the near future? Johnson wants nothing to do with Detroit at this time because, as he puts it: "It's too emotional - we'd get smashed into the earth!" The inventor is equally reluctant to make predictions about other applications as well, mainly because he just wants time to perfect his ideas, and hopefully, get the scientific establishment to at least consider his unorthodox ideas with a more open mind.

For example, Johnson argues that the magnetic forces in a permanent magnet represent super-conductance that is akin to phenomena normally associated only with extremely cold superconducting systems. He argues that a magnet is a room temperature superconducting system because the electron flow does not cease, and because this electron flow can be made to do work. For those who scoff at the idea that permanent magnets do work, Johnson has an answer: "You come along with a magnet and pick up a piece of iron, then some physicist says you didn't do any work because you used that magnet. But you moved a mass through a distance, right? That's work that requires energy. Or you can hold one magnet in the air indefinitely by positioning it over another magnet with like poles facing. The physicist will argue that because it involves magnetic repulsion, no work is done. Yet if you support the same object with air, they will agree in a minute that work is done!"

There's no doubt in Johnson's mind that he has succeeded in extracting usable energy from the atoms of permanent magnets, but does that imply that the electron spins and associated phenomena that he thinks provide this power will eventually be used up? Johnson makes no pretense of knowing the answer: "I didn't start the electron spins, and I don't know any way to stop them - do you? They may eventually stop, but that is not my problem."

Johnson still has many practical problems to solve to perfect his invention. However, his greatest challenge may be to win general acceptance of his ideas by an obviously nervous scientific community in which many physicists remain compulsive about defending the law of conservation of energy without ever wondering whether that "law" really needs defending.

The dilemma facing Johnson is not really his dilemma, but rather that of other scientists who had observed his prototypes. The devices obviously do work. Yet, the textbooks say they shouldn't work. All that Johnson is really saying to the scientific community is this: Here is a phenomenon which seems to contradict some of our traditional beliefs. For all our sakes, let's not dismiss it outright, but take the time to understand the complex forces at work here.

The idea of a permanent magnet self-powering motor has been pursued by many inventors, and successfully by some. The Steorn group in Ireland had and has such a working motor, but did not understand the vacuum interaction occurring in the asymmetry of the motor. They also did not understand that the local vacuum (and its interaction with one's motor or other gadget using energy from the vacuum) can vary from place to place, as pointed out to Frank Golden and myself by Professor William Tiller, former head of the Materials Science Department at Stanford. Tiller had run into that specific effect and so had become aware that one often "grew an archetype or conditioned form" in that local vacuum exchange with one's system. He personally had a system to demonstrate this "growth" of Tiller's "archetype" for his special detector.

Golden ran into that effect with a Golden over-unity machine which, when relocated several hundred miles, suddenly did not produce COP>1.0 in the new location. They were thunderstruck until Tiller explained it to them.

The demonstration intended for Steorn held so close to the original development site was catastrophic to them.

All such successful self-powering asymmetric permanent magnet motors have been ruthlessly suppressed to date. The latest highly suspicious "capture" of a legitimate procedure to make asymmetric permanent magnets – and thus suppress the process from use by our citizens – was by our own U.S. Department of Energy in 2001. Apparently, the DoE has also kept this fact – the availability of the mass production of asymmetric permanent magnets with high energy products, and thus, the easy assembling of self-powering asymmetric PM motors and motor generators – is completely unknown to the President, the Congress, and the American people. There's an article on our website with the three international patent numbers of interest. In the 2001 patent where DoE "funded a grant that produced the process again, after it had already been invented", you can see DoE's statement essentially assuming control of the process.

The idea of an asymmetric permanent magnet motor is certainly not new. The late Howard Johnson, for example, pursued it arduously for decades. He succeeded at least once in achieving such a demonstrable self-powering model which I personally tested for several hours, and shortly thereafter, that specific demonstrator was promptly stolen from him by a very professional, pre-dawn break-in to his laboratory in Virginia. The only thing stolen was that specific successful self-powering little Johnson demonstrator.

John Bedini is certainly capable of making such a motor, and he has done so in the past using his own special methods. Presently, he is vigorously trying to get his advanced battery chargers into production and onto the market. John and his few assistants are trying to do the work of at least two dozen men, and I praise him for making the most valiant effort I've ever seen to get highly effective systems out there onto the world market.

DeGeus suffered a "mysterious" fatal heart attack when he parked his car in the Douglas International Airport on his way to the Netherlands to sign a contract and receive very substantial funding to get his successful "self-powering battery" on the world market and give us self-powering electrical cars. You can take tourmaline and reduce it to crystals, and get a strange effect. Each crystal is like a little battery, with a positive charge on one side and a negative charge on the other. In short, it is a dipole. You can use a tiny conductor across that "crystal battery" to power a tiny load. Eerily, the crystal will continue to power the load and to also maintain its own source dipolarity. In short, it will continue freely extracting the EM energy from the local vacuum via the self-maintained asymmetry of its source dipole, and thus, freely powering its load.

Make a plane of such crystals-oriented parallels, and you will increase the available current. Increase the "depth" by adding other planes behind it, and you also increase the voltage.

If one gains good control of the process, and can get to other materials by finding how to elicit the self-sustaining dipole effect, then the self-powering battery for self-powering electric automobiles becomes readily possible.

The water-gas process is also again very extant, and several inventors have working model systems that easily produce COP>1.0. Kanzius' process is one such example, and his related cancer cure method has passed initial animal testing with 100% success (100% of the cancers in the lab animals were simply and easily cured, without all the harmful radiation and without unnecessary damage to the animals).

For PM self-powering motors, you could, in theory, also use "edge switching" (switch the magnetic field in just the edge of a bar magnet to just a minor depth), which can also be used to make the back mmf region and the forward mmf region asymmetric.

The point is this: It isn't the power that we externally input into a coil in a permanent magnet motor, but a coil judicially placed in the back mmf section that allows us to break the symmetry of the back mmf region. That asymmetry – once achieved by any means – is what actually powers the system! The only thing that inputting our paid-for energy to the coil does is *produce the necessary asymmetry*.

The materials, motor and assembly of the motor system do not care in the least how we obtain that asymmetry in the first place. Get it anyway you wish, and then the motor will power itself and a load, taking its input energy directly from the seething virtual state vacuum via that broken symmetry.

Every joule of EM energy in the universe and in all of our circuits is already extracted directly from the seething virtual state vacuum by the asymmetry of the source dipole. It is NOT the cranking of the shaft of the generator that powers the circuit! All that the cranking input does is continually restore the source dipole inside the generator that our electrical engineering symmetrical circuits and systems keep destroying faster than they power their loads!

As Tesla (who had discovered energy from the "active medium" circuits three decades earlier) stated in 1919:

"The Hertz wave theory of wireless transmission may be kept up for a while, but I do not hesitate to say that in a short time, it will be recognized as one of the most

remarkable and inexplicable aberrations of the scientific mind which has ever been recorded in history." [Nikola Tesla, "The True Wireless," Electrical Experimenter, May 1919].

The 1892, symmetrization of the Heaviside equations (erroneously taught to our EEs as "Maxwell's theory") was specifically done to stop Tesla cold and prevent him from giving the world free energy systems, taking their energy directly from the "active medium". Quoting Tesla earlier:

"Ere many generations pass, our machinery will be driven by a power obtainable at any point in the universe. This idea is not novel... We find it in the delightful myth of Antheus, who derives power from the earth; we find it among the subtle speculations of one of your splendid mathematicians...Throughout space there is energy. Is this energy static or kinetic? If static, our hopes are in vain; if kinetic – and this we know it is, for certain – then it is a mere question of time when men will succeed in attaching their machinery to the very wheelwork of nature." [Nikola Tesla, in a speech in New York to the American Institute of Electrical Engineers, 1891. Quoted from back cover of his biography, Margaret Cheney, Tesla: Man Out of Time].

What we should be doing in our vaunted scientific community is coming down on our entire electrical engineering establishment, and brute-forcing them to go back and *remove that insane symmetrization of the Heaviside equations* that Lorentz shoved on them arbitrarily in 1892.

In short, turn all our young doctoral candidates and post-doctorates loose on the problem of developing and using practical asymmetrical Maxwellian systems, and in two years, there will never again be another energy crisis anywhere on this planet.

Advanced physics (Nobelist Lee himself stated it) already tells us (and has done so for 50 years) that when we do have a broken symmetry, then in that broken symmetry, *something virtual has been and is being converted to something observable*.

In short, any broken symmetry, such as a dipole or a charge with its surrounding polarized vacuum, is ALREADY extracting and integrating real EM energy from the seething virtual state vacuum energy it continually absorbs, and then pours out this free EM energy (real observable photons) freely and steadily.

It is astoundingly simple to evoke an everlasting *free electromagnetic energy wind* from the active vacuum. For example, just place a charged capacitor on a permanent magnet so that the E-field of the cap is at right angles (90°) to the H-field of the magnet. That silly 2-dollar gadget will then sit there indefinitely and freely pour out real Poynting EM energy flow S, given by the simple equation S = E X H. *That equation is verifiable in every EM textbook in the world, over in the Poynting energy flow theory section*.

To sum it up, we can easily get real EM energy flowing freely from the local inexhaustible virtual state vacuum, and it will do so indefinitely. There is no free energy problem! Instead, there is the problem of building and using properly ASYMMETRIC collection and dissipation systems to power our loads.

To make a proper "EM windmill" to intercept and collect some of that easily-achieved and real EM energy flow, and also to SEPARATELY dissipate the collected EM energy in the loads to power them, an ASYMMETRIC system is required.

Thus, our inane electrical engineering profession – since 1892 at its very beginning – has all our electrical engineers building a SYMMETRICAL windmill, and then having us pay to burn fuel, etc. to input energy to the "windmill" that does nothing but produce the asymmetry.

We've even had group theory in our universities since 1870! Yet, if you ask the average EE professor, "What were and are the ramifications of Lorentz's symmetrization of the Heaviside equations in 1892, and the continuing symmetrization of these equations today as the "electrical engineering model", the professor won't have the foggiest notion of what you have asked him or what you have steered him to.

As long as our engineers inanely continue to build only SYMMETRICAL Maxwellian/Heaviside systems, then we shall have to continue to feed in energy that we must pay for ourselves to produce the necessary asymmetry.

The energy crisis in this world – and the tragic deaths of hundreds of millions of people worldwide because of their abject poverty due to a lack of cheap energy for a modern national economy – rests directly on the shoulders of our scientific community, and specifically on the shoulders of our electrical engineering community and profession.

Anyway, with the present rapidly escalating oil prices and gasoline prices, the U.S. seems headed rather quickly toward a great economic collapse. If the Middle East explodes (which is likely any time now), that will immediately chop off the rest of the region's oil from the U.S. and Western Europe. Then, we will not be discussing the price of gasoline at our pumps. Instead, we will be asking if there is a pump anywhere that *has any gasoline left*.

If the big diesel trucks stop for four or five days, the food in our cities will also be depleted (it will not be replenished daily by the big trucks). In that case, there will be famine, rioting, looting, and chaos in our cities, requiring the immediate application of very rigorous martial law.

Watch for it! Unless our scientific community wakes up from its century-long sleep soon, and gets its head out of the sand in its ostrich position that it's been in since 1892, then that's exactly what we are headed for.

Can the asymmetric PM motor generator that you described in the 01 July 2007 Technical Paper be correctly called a Bearden Motor? In other words, is it your conception? Will a piece of iron placed on one side of a magnet to reduce the field strength on that side produce the desired asymmetry?

There are many different formulations of NdFeB that are used for neodymium magnets, each with different magnetic properties and field strengths. Can neo-magnets be used if the layers consist of different formulations of NdFeB, each successive layer having less magnetic field strength (BHmax)? In this case, different formulations are compacted as thin strips in an aligning magnetic field, so that all the domains are pointing north/south. Afterward, the strips undergo a final compaction in a die, also within an aligning magnetic field. The resulting layered magnetic material is then sintered, and finally, magnetized to saturation.

Combining the Magnetic Engine with an Air Cycle Machine

The magnetic air car is a vehicle which does not require fuel or a charge to operate. The objective is to develop a motor (approx. 200 lbs.) to fit inside a Factory 5 chassis (400 lbs.) which will be able to manufacture and recycle its own air. In order to accomplish this, MACx will need magnetic motor technology with capabilities of producing 5-10 hp to maintain the air compressor pump and keep the battery charged. The magnetic motor technology will need controllers and electronics to operate the motor.

Motor Development Project

Development of the magnetic motor will involve the following:

- Positioning of extremely strong neodymium magnets in a pattern that allows virtually zero torque to rotate the motor shaft without power.
- Development of a microcontroller-based control system for the motor that allows maximum power from the coils driving the magnets. The system will have dash control system reflecting output power, fuel level, speed, torque, etc.
- Maximize the output power-to-weight ratio. (Goal of 1 hp per pound)
- Maximize efficiency. (Goal: >95% efficiency)
- Build a complete working prototype.

This technology will revolutionize our access to energy.

The energy machine is proof that the strength of the motor's magnetic field is dependent upon the VOLTAGE --- NOT the CURRENT!

For the first time, Joseph Newman demonstrated an energy machine that turned a 1,650-lb rotary and powers a 375-lb positive displacement pump using LESS THAN 20 WATTS!

In his video demonstration, the energy machine produced phenomenal speed and torque, and continuously pumps water using LESS THAN 20 WATTS!

LESS THAN 20 WATTS!!!

Conventional technology would say that this is not possible. Yet, the Newman Energy Machine can accomplish that ... and much more. In fact, the technology can be used to provide abundant, non-polluting, inexpensive electromagnetic energy to power the world.

No more water or energy shortages - anywhere on the planet.
No more fossil-fuel-related pollution.
No more skyrocketing energy costs.
No more global warming caused by the use of hydrocarbon fuels.
No more nuclear reactor wastes.
No more coal-burning power plants.

This technology will provide abundant, inexpensive, non-polluting energy to power homes, factories, automobiles, ships, planes, appliances, etc. In short, the entire world can utilize this revolutionary technology.

I, Joseph Newman repeatedly stated that mass means NOTHING if it does not have a MECHANICAL UNDERSTANDING . The equation for horsepower of rotating motor shaft so PROVES - "ft. lbs X rpm divided by 5252 = HP". Yet, that equation does not designate the TIME of said horsepower. Constant of mass - rpm implies horsepower per minute.

In the late 1700's, James Watt, who established the HP as a unit of measurement, specifically stated that such a measurement was dependent upon TIME: "33,000 lbs 1 ft. per MINUTE."

Accordingly, the HP of a rotating shaft must be measured in HP per minute of rotating shaft, or, in shorthand, HP of rotating shaft per SECOND. ["550 lbs 1 ft. per second."]

I show you now a simple MECHANICAL explanation of HP per second of TRUTH: Take a shaft and attach it to the center of 1 ft radius [2 ft. diameter] pulley. Now imagine that you rotate the shaft of one full rotation of 360 degrees in one SECOND. Then "you" have lifted 550 lbs 6.283 ft in one second! ["1 ft. radius gives 6.283 ft in circumference"] In other words, 6.283 HP per second! Per rotation of shaft! Multiply by [60 rpm] 60 SECONDS, and that equals 376.98 HP per MINUTE!

The above equation is WRONG! Vague conventional HP equation gives only 6.283 HP per MINUTE!

The TRUTH of MECHANICAL explanation that I've given has been applied to rotating the 400-lb. shaft of the magnetic rotary of my revolutionary motor that will shock the world in the immediate future. IT WILL PROVE WHO IS THE MASTER OF MY REVOLUTIONARY ENERGY INVENTION!

Fact: Conventional equation must be changed to the following:
HP of rotating shaft of motor:
HP = ft lbs X RPM divided by 5252 X 60 = HP per MINUTE.

The demonstration held at the National Press Club in Washington, D.C. on March 28, 2005 went as predicted.

The 100-lb. Newman Energy Machine contained a 70-lb. coil with a rotor that operated at a HIGH RATE OF RPM with only MINUTE SPARKING as a result of the back-emf produced by the machine while it operates. The energy machine operated on a low voltage of only 30 volts consisting of five 6-volt batteries each comprised of four 1.5 volt battery units -- with a current equivalent to that of any one of those 1.5 volt battery units.

Such important technical developments enable mass production of this technology to go forward!

I also wish to speak about another subject to the good people who support my life's work for humanity:

During the news conference, I had to physically remove an individual from the conference room who, along with his physicist cohort, had come to the news conference to heckle me and my work with no interest in understanding the importance of my energy machine technology.

I will now explain why I did that. I am 68 years of age and I have experienced a LOT during the course of my life. When I saw the mannerisms of those two individuals towards my presentation and heard their sarcastically-toned "forked" tongues, I became enraged. I was enraged for all humanity and for all creative minds that have ever been subjected to such insensitive and uncaring attitudes.

For the past 40 years, I have clearly witnessed that many so-called scientific intellectuals in high positions of influence are FRAUDS. Why? Because they have the mentality of a parrot. They have simply memorized the great contributions of important scientific innovators of the past and can quote those contributions like a parrot. Yet at the same time, they have absolutely no true understanding or curiosity regarding the subjects they quote.

I tried for years to get them to debate me and they always refused. They said the "burden of proof is on me."

I then put forth the absolute proof of a Unified Mechanical Field Theory, followed by the Energy Machine. But the mental cowards have still refused to debate me.

I do however give credit to more than 40 brave scientific individuals (physicists, electrical engineers, nuclear engineers, mechanical engineers, etc.) who did come forth and endorse the magnitude of my Mechanical Theory and the world importance for humanity resulting from the development of my Energy Machine.

I have total contempt for the mental cowards and parrots in the scientific community whose attitudes impede the progress of civilization for all humanity. And I give deep love and thanks to the courage of John Denver and others who have helped the Energy Machine to go forward.

Sincerely,
- Joseph W. Newman

"In speaking of the Energy of the field, however, I wish to be understood literally. All energy is the same as mechanical energy, whether it exists in the form of motion or in that of elasticity, or in any other form. ***The energy in electromagnetic phenomena is mechanical energy****."*

- JAMES CLERK MAXWELL

"The theory I propose may ... be called a Theory of the Electromagnetic Field because it has to do with the space in the neighborhood of the electric or magnetic bodies, AND IT MAY BE CALLED A DYNAMICAL THEORY, BECAUSE IT ASSUMES THAT IN THAT SPACE THERE IS MATTER IN MOTION, BY WHICH THE OBSERVED ELECTROMAGNETIC PHENOMENA ARE PRODUCED."

- JAMES CLERK MAXWELL

As a young man, inventor Joseph Newman did question "certain things (that) have unequivocally been observed to be so" --- after all, this is one of the principal ways in which new scientific understandings of the universe can occur, by questioning "the accepted explanations" of a given paradigm.

Joseph Newman was raised as a "hard-nosed country boy" (to employ the colloquialism), and as such, he did *not* believe in getting "something for nothing"..... sometimes referred to as "perpetual motion." His mind instinctively rebelled against such a notion.

Yet, when he observed that one's motion (slow) of one's hand holding a conductor wire and moving it through a magnetic field (in a particular orientation) produces an electric field (very fast) along the same conductor, this superficially appeared to him to be a case of one obtaining "something (very fast) for nothing (slow)." He refused to believe this could be the case since he did not believe in "perpetual motion." He chose to investigate further.

Through his studies, it became obvious that "something" was ALREADY moving (very fast) in that (electro)magnetic field, and this (very fast) "something" was simply being DEFLECTED onto the conductor as it moved (slowly) through the magnetic field.

This conclusion answered his first question. If he had the attitude of some conventionally-educated scientists, he would just leave it at that and be satisfied.

He opted to question further:
What was the precise nature of this "something"?
Also, there's this question: does this "something"
have a MECHANICAL nature?

Well, two of the greatest minds of the last 200 years believed that it DID.

Their names:
Michael Faraday (the mechanical genius) and
James Clerk Maxwell (the mathematical genius)

I believe that Joseph Newman would prefer to "stand on the shoulders" of these two great men rather than operationally accept the status quo and "lounge on the postulations of some conventionally-educated, present-day scientists."

[I should add that there are certainly those in the teaching profession who are indeed exceptions to the above -- those who are truly curious, rational, and intellectually honest.]

And what did these two men state?
Michael Faraday wrote:

"I cannot conceive curved lines of force without the conditions of a PHYSICAL existence in that intermediate space."

- MICHAEL FARADAY

Faraday also stated:

"How few understand the PHYSICAL lines of force! They will not see them, yet all the researches on the subject tend to confirm the views I put forth many years since. Thompson of Glasgow seems almost the only one who acknowledges them. He is perhaps the nearest to understanding what I meant. I am content to wait convinced as I am of the truth of my views."

- MICHAEL FARADAY

And what does James Clerk Maxwell have to say on this subject?

"In speaking of the energy of the field, however, I wish to be understood LITERALLY. All energy is the same as MECHANICAL ENERGY, whether it exists in the form of motion or in that of elasticity, or in any other form. The energy in electromagnetic phenomena is MECHANICAL energy."

- JAMES CLERK MAXWELL

That's a pretty strong, rather unequivocal statement on the part of James Clerk Maxwell.

Just to ensure that no one might later misinterpret his remarks, he added, with EMPHASIS:

"I WISH TO BE UNDERSTOOD LITERALLY"

Literally.

Well, "literally", in my book, means LITERALLY.

And what is James Clerk Maxwell *literally* saying about the nature of the energy of the field…?

…that ALL --- ALL --- ENERGY is the SAME as MECHANICAL ENERGY……. that all such energy in ELECTROMAGNETIC

PHENOMENA is MECHANICAL ENERGY.

As physicist Dr. Roger Hastings (who endorsed Joseph Newman's work) wrote regarding Maxwell's emphasis on the MECHANICAL nature of energy:

"This is stated, in no uncertain terms, in Maxwell's book *A Dynamical Theory of the Electromagnetic Field*. In fact, Maxwell used a dynamical model to derive his famous equations. This fact has been all but lost in current books on electromagnetic theory. The quantity which Maxwell called 'electromagnetic momentum' is now referred to as the 'vector potential.'"

And in a reference to the work of Richard Feynman, physicist Robert Matherne (who also endorsed Joseph Newman's work) wrote:

"While reading my *Commemorative Issue of The Feynman Lectures on Physics by Feynman, Leighton, and Sands,* published by the Addison Wesley Publishing Company, I encountered a passage on relativistic momentum considerations by Richard Feynman that sounded so similar to what Joseph Newman said in his *Energy Machine* book that I wanted to share it. Here's the operant quotation from page 10-9 of Volume I:

'One of the propositions of Newton was that interactions at a distance are instantaneous. It turns out that such is not the case; in situations involving electrical forces, for instance, if an electrical charge at one location is suddenly moved, the effects on another charge, at another place, do not appear instantaneously -- there is a little delay.

It takes time for the influence to cross the intervening distance, which it does at 186,000 miles per second. In that tiny time, the momentum of the particles is not conserved. Of course, after the second charge has felt the effect of the first one and all is settled, the momentum equation will become balanced, but during that small interval, momentum is not conserved. We represent this by saying that during this interval, there is another kind of momentum besides that of the particle (mv), and that is momentum in the electromagnetic field. If we add the field momentum to the momentum of the particle, then momentum is conserved at any moment all the time. The fact that the electromagnetic field can possess momentum and energy MAKES THAT FIELD VERY REAL, and so, for better understanding, the original idea that there are just the forces between particles has to be modified to fit the idea that the particles make a field, and a field acts on another particle, and that THE FIELD ITSELF HAS SUCH FAMILIAR PROPERTIES AS ENERGY CONTENT AND MOMENTUM, JUST AS PARTICLES CAN HAVE.' [Emphasis added.]

"From the first edition of *The Energy Machine of Joseph Newman*, page 20, I quote:

'The FACTS above clearly indicate that the magnetic field consists of GYROSCOPIC TYPE MASSERGIES which are the mechanical essence of $E = MC^2$ and represent an orderly flow of kinetic energy.' [Emphasis added.]

"Note the similarity between the Feynman and Newman descriptions of the electromagnetic field: both say the field has energy content and momentum. Feynman, however, pulls his punches by inserting the scientific qualification, '...just as particles can have'; while Newman says directly the field consists of 'particles in motion.' PARTICLES in motion, by definition, have momentum, and the electromagnetic field, according to Nobel Laureate Feynman, in a stretching of the very definition of momentum, has momentum. Since no one has ever seen either an electric field nor gyroscopic particles (nor will, presumably, they EVER, since seeing requires the objects be larger than a wavelength of light, which they definitely are not), both are legitimate ways of talking about a reality that we cannot observe directly. Feynman's way of talking used the concept of field in the accepted scientific say, up until now. Here's a later quote by Feynman from the same page 10-9:

'To take another example --- an electromagnetic field has waves, which we call light; it turns out that light also carries momentum with it, so when light impinges on a object, it carries in a certain amount of momentum per second; this is equivalent to a force, because if the illuminated object is picking up a certain amount of momentum per second, its momentum is changing and the situation is exactly the same as if there were a force on it. Light can exert pressure by bombarding an object; this pressure is very small, but with sufficiently delicate apparatus, it is measurable.'

> *"If the momentum of the electromagnetic fields or particles of the sun's rays can be used, as some scientists have proposed, to build a solar sail to power future space ships throughout the solar system, perhaps the momentum of the electromagnetic fields surrounding large coils of wire can be used to provide an efficient source of electrical power throughout the same solar system without the use of fossil or radioactive fuels. This is the promise of Joseph Newman's revolutionary energy machine and its associated theory."*
>
> **- Robert Joseph Matherne, Physicist**

But Joseph Newman isn't the only one describing such fields as "matter in motion." James Clerk Maxwell 'beat him to the punch':

Maxwell also wrote:

> *"The Theory I propose may ... be called a Theory of the Electromagnetic Field because it has to do with the space in the neighborhood of the electric or magnetic bodies, AND IT MAY BE CALLED A DYNAMICAL THEORY, BECAUSE IT ASSUMES THAT IN THAT SPACE THERE IS MATTER IN MOTION, BY WHICH THE OBSERVED ELECTROMAGNETIC PHENOMENA ARE PRODUCED."*
>
> **- JAMES CLERK MAXWELL (Emphasis added.)**

These are pretty strong statements on Maxwell's part.... and he did not intend for the scientific establishment, which came after him, to ignore his words.

Operationally speaking, they have.

For the historical record:

I should state that Joseph Newman's conclusion about the fundamental mechanical nature of (electro) magnetic fields was reached years before he became aware of Maxwell's position on this matter. For Joseph Newman, Maxwell's strong statements served as corroboration from a mathematical genius of the rightness of Joseph Newman's paradigm.

All right.... to continue the discussion....
...... MATTER IN MOTION....

Joseph Newman took the concept of "MATTER IN MOTION" and integrated it with his earlier observations regarding Faraday's Generator, i.e., that "something" was ALREADY moving (very fast) in that magnetic field, and this (very fast) "something" was simply being DEFLECTED onto the conductor as it moved (slowly) through the magnetic field.

Next fundamental question:
What is the operational, mechanical nature/behavior of this "something"?

To say that a magnetic field consists of "matter in motion" is an advance in our fundamental understanding of such a field. Unfortunately, it doesn't tell us much about the real, physical characteristics of this "matter in motion."

Perhaps if one was to fundamentally and deeply understand the mechanical behavior of these "matter(s) in motion," then one might be able to even conceive a new way of constructing motors/generators that would enable one to more efficiently harness the pre-existing, real, kinetic, physical, MECHANICAL MOTION of these "somethings."

It was approximately at this point in his studies that Joseph Newman came across Fleming's Rule.

Now, what does Fleming's Rule state?

Fleming's Rule states:
"If the fingers of the right hand are placed around a current-carrying wire so that the thumb points in the direction of the conventional current, the fingers will point in the direction of the magnetic field."

Great. That is a convenient "rule." It is a credible-by-observation "rule."

Great. Electrical engineers are aware (presumably) of that rule.

Yet, accepting the limited paradigm of the rule does not answer the more fundamental question:

What MECHANICALLY --- physically --- causes the rule to be true?

That is the most fundamental question that occupied Joseph Newman's mind during the 1960s. He sought an honest answer to that more fundamental question as to what is the MECHANICAL cause of Fleming's Rule. He could not find an answer to that question in any of the books that he read on electromagnetics and electrical engineering. Yet, he believed it was absolutely essential to answer that question, if he was to progress in a deeper understanding of the fundamental MECHANICAL nature of (electro) magnetic fields [per Faraday's and Maxwell's statements above].

It was about this time in his studies that he began studying the GYROSCOPE. He had heard that the "gyroscope" was a stabilizer, but did not know much more about its nature. As a hobby, he began mechanically working with gyroscopes and children's bicycles (he converted the bicycle of a neighbor's son to a clutch system to enable it to perform "wheelies" better) --- so he began to acquire a first-hand understanding of the behavior of GYROSCOPES.

At one point in this process, he *instantly* had an answer to the question that had "boggled his mind" for so long. He *instantly* knew that those "somethings" in that magnetic field surrounding Faraday's Generator individually behaved as miniature "GYROSCOPES."

Many related concepts then began to "click" in his mind. It was then that his real insights into the mechanical nature of (electro) magnetic fields began to generate a new paradigm concerning such fields.

As he explains in detail in chapter two of his book:

[Note: I will describe only in words what is graphically and with words depicted step-by-step in the book.]

> 1-A) If one has north and south ends of magnets facing one another with a gap between them (as in Faraday's Generator), if one moves a conductor DOWN at a right angle to the magnetic field, the CURRENT FLOW MOVES LEFT.

1-B) If one applies a downward force to the axis of a spinning gyroscope, it will pivot at a right angle to the force (in this case, of Figure 11-A1), and it will PIVOT LEFT. Now imagine that this gyroscope has a forward motion at the speed of light as well as spins at the speed of light.

2-A) If one has north and south ends of magnets facing one another with a gap between them (as in Faraday's Generator), if one moves a conductor UP at a right angle to the magnetic field, the CURRENT FLOW MOVES RIGHT. (Figure 11-B)

2-B) If one applies an upward force to the axis of a spinning gyroscope, it will pivot at a right angle to the force (in this case of Figure 11-B1), and it will PIVOT RIGHT and opposite to the case of 1-B above. Now imagine that this gyroscope has a forward motion at the speed of light as well as spins at the speed of light.

3-A) Now flip the above-described magnets over 180 degrees: If one has a south and north magnet facing one another with a gap between them (as in Faraday's Generator), if one moves a conductor DOWN at a right angle to the magnetic field, the CURRENT FLOW MOVES RIGHT and opposite to that of case 1-A, even though the force direction is the same. (Figure 11-C)

3-B) Now flip the above-described gyroscope over 180 degrees: If one applies a downward force to the axis of a spinning gyroscope, it will pivot at a right angle to the force (in this case of Figure 11-C1), and it will PIVOT RIGHT and opposite to the case of 1-B above, even though the force direction is the same. Now imagine that this gyroscope has a forward motion at the speed of light as well as spins at the speed of light.

4-A) With the magnets in the same position as 3-A, if one has a south and north magnet facing one another with a gap between them (as in Faraday's Generator), if one moves a conductor UP at a right angle to the magnetic field, the CURRENT FLOW MOVES LEFT and opposite to that of case 2-A, even though the force direction is the same. (Figure 11-D)

4-B) With the gyroscope in the same position as 3-B, if one applies an upward force to the axis of a spinning gyroscope, it will pivot at a right angle to the force (in this case of Figure 11-D1), and it will PIVOT LEFT and opposite to the case of 2-B above, even though the force direction is the same. Now imagine that this gyroscope has a forward motion at the speed of light as well as spins at the speed of light.

5-A) With the magnets in the same position as 3-A, move the conductor vigorously "up" and "down" through the magnetic field, MAINTAINING THE CONDUCTOR FORCE PARALLEL TO THE MAGNETIC LINES OF FORCE AND *NO* CURRENT FLOW IN THE CONDUCTOR WILL RESULT.

5-B) Apply a vigorous "up" and "down" force parallel to the axis of the spinning gyroscope. Regardless of how energetically the force is applied, as long as the force remains parallel, the gyroscope will not pivot, even though it has a forward motion at the speed of light.

THE ANALOGY OF THE ABOVE TWO EXAMPLES IS SCIENTIFICALLY EXACT!

Once Joseph Newman began to understand that the "mechanically moving somethings" in a magnetic field each possessed a GYROSCOPIC SPIN, then many other insights began to fall into place.

It was at this time that Joseph Newman also began to understand and mechanically explain:

1. Why, in a mechanical sense, does a magnet attract and repel other magnets?
2. Why, in a mechanical sense, do electric charges attract and repel?
3. What is the energy in a magnetic field and what is its source?

4. Did the energy used in creating a permanent magnet have any bearing upon the strength or energy contained within a magnetic field emitted from the permanent magnet once it was made?

As Joseph Newman writes in chapter two of his book,

"In the early part of 1965, I eagerly researched the known facts concerning the creation of a permanent magnet. Because I instinctively knew that if the strength of a magnetic field was solely relative to the energy input, I knew that I was incorrect. However, if the strength of the magnetic field was INDEPENDENT of the energy input, then I would be even more assured that I was correct.

"Upon examining the known facts concerning permanent magnets, I once again knew that I WAS correct."

It took Joseph Newman over 15 years to develop and refine his Theory and Technical Process. It was not until these many years passed did he even construct his first crude, hand-built prototype to demonstrate the operability of his Technical Process. Ironically, Joseph Newman *knew* that his invention would work and demonstrate the nature of his Technical Process. The prototype(s) were needed to convince the rest of the world. I should add that Joseph Newman considers his Theory to be <u>far more important</u> than the application.

This attitude on his part is analogous to Einstein and his Theory of Relativity.

Earlier in the 20th century, for a number of years, there was no "proof" of Einstein's Theory -- it was just a challenging (and sometimes, upsetting to others) idea in Einstein's head.

It was not until the Ecliptic Expedition of 1919 (chaired by Arthur Eddington) that Einstein's Theory was corroborated. During 1919, a journalist on hand who was following the Expedition cornered Einstein and asked him, *"Dr. Einstein, your entire Theory hangs in the balance of this Expedition. Aren't you a bit concerned as to its outcome?"*

Characteristically, Einstein replied, *"Oh no. I'm not worried. I know my Theory is correct. The Expedition is only needed to convince the rest of the world."*

My introduction to Joseph Newman's work nearly 20 years ago was through an understanding of his revolutionary explanation for Magnetic Attraction and Repulsion. It was at that point that I became convinced of the validity of his theory.

In essence, a MAGNETIC FIELD mechanically consists of "negative" and "positive" gyroscopic particles having opposite spins and simultaneously traveling in OPPOSITE directions, e.g., the concentric "shells of force" surrounding a bar magnet.

An ELECTRIC CURRENT mechanically consists of "negative" and "positive" gyroscopic particles having opposite spins and traveling in the SAME direction down a conductor wire ---similar to light traveling in space through a medium.

> *"In speaking of the energy of the field, however, I wish to be understood literally. All energy is the same as mechanical energy, whether it exists in the form of motion or in that of elasticity, or in any other form. The energy in electromagnetic phenomena is mechanical energy."*
>
> \- JAMES CLERK MAXWELL

"Eight slightly-used penlight batteries were connected (in series) to a conventional electric motor. The conventional motor operated for 1 minute and 15 seconds before stopping."

"These same penlight batteries were then connected to a portable model of Joseph Newman's motor/generator. This motor/generator proceeded to run, and a 90-pound magnet continuously rotated for 1 HOUR and 15 MINUTES, at which time the WWL-TV engineers disconnected the batteries because of a lack of time on the part of the film crew, which had to return to the TV station to produce the evening broadcast.

"Afterward, the above-described penlight batteries were RECONNECTED to the ORIGINAL conventional electric motor. THIS motor operated for TWO minutes and 28 seconds before the conventional motor stopped. This is nearly

TWICE as long as the first time (above) --- using batteries that are not supposed to be rechargeable!"

> *"I cannot conceive curved lines of force without the conditions of a PHYSICAL existence in that intermediate space."*
>
> - MICHAEL FARADAY

Applicant [Joseph Newman] used a permanent magnet D.C. motor as a generator. He demonstrated that the resistance of the copper windings of the generator was only 3 ohms (Ω).

Therefore, he demonstrated the so-called work load would be nil if the two leads from the generator were connected, and the generator shaft were then rotated by hand (pulling a cord wrapped around a 1.5" diameter pulley attached to shaft of generator). I was asked to then pull the cord lightly once, and then briskly. I immediately experienced noticeable resistive force the harder I pulled the cord, although there was no conventional work load hooked in the system. Applicant mechanically explained these results by his teachings of gyroscopic particles; that when the atoms of the rotating coils of the generator hit the gyroscopic particles (at some degree of a right angle) which were being emitted from the atoms of the permanent magnets in the generator, that the gyroscopic particles then went down the length of copper wire coils (but that their gyroscopic spin would then be at some degree of a right angle to the balance of the spin of the gyroscopic particles still moving in the magnetic field from the permanent magnets). Therefore, when the leads were hooked together, this allowed the gyroscopic particles to try to re-enter the influences of said gyroscopic particles of said permanent magnets, but that their spin would be at some degree of a right angle to one another, and therefore, they try to push away from each other, resulting in the coils of [the] generator having resistance to rotation. Additionally, this effect was multiplied the faster you turned the coils, because it would cause more gyro-particles to be released from said permanent magnetic field, resulting in an ACCUMULATIVE EFFECT of gyro-particles in the closed system (coils) trying to re-enter the influence of gyro-particles moving in said magnetic field of said permanent magnets, and therefore, would always more vigorously resist your acceleration of the coil and its shaft of the generator, although there was no conventional work load placed in the system.

Later, applicant [Joseph Newman] attached work load of six small miniature motors in series to the two leads of the generator (the resistance or conventional work load placed in the system was then considerably greater). Applicant now asked me to again pull the cord as I had previously done. I must say I was amazed to find that the resistance to me rotating the shaft of the generator was dramatically reduced, although all six motors ran when I pulled the cord and work was obviously being produced.

If, as the prior art so persistently teaches, the electrical energy produced by a generator or battery is used up in the work load which the system operates, then why should I observe these results? As applicant [Joseph Newman] pointed out, the conventional words "shorted out" does not mechanically explain the results. However, his teachings and instructions do mechanically explain the results.

Applicant [Joseph Newman] explained that when the work load of six motors in series was placed in the system, the same number of gyro-particles were released from the field of the permanent magnet by the coils of the generator (relative to equal speeds of rotation of shaft), but now, because of the work load (resistance), they could not easily re-enter into the gyro-particles of said magnetic field of said permanent magnets, and therefore, resulted in a hydraulic effect back to their source and throughout the work load of the motors, resulting in the motors rotating and producing obvious work, and yet, causing me less energy input. Applicant [Joseph Newman] pointed out that once the gyro-particles managed to get through the work load of the motors, then they would still go back into the generator and cause me to experience a resistive force (as a result of prior explanation). Therefore, based off applicant's teachings, one does not want the electric current to get back to the source of its beginnings, whether battery or generator. Also, contrary to prior art teachings, the electrical current is NOT used up in the work load! Instead, the input of a work load causes LESS destruction of a battery or LESS energy input into a generator, exactly as applicant teaches in his specification. Applicant also points out that the prior art teaches [that] copper is "nonmagnetic", and contrary to this, applicant teaches throughout the specification concerning Figs. 5 and 6 that copper is extremely magnetic, so much so that an individual is easily fooled into thinking copper is nonmagnetic, simply because the magnetic field will disappear so quickly when the current is turned off.

Applicant [Joseph Newman] also vigorously teaches that the energy in the field of force of any type of magnet is the energy which makes up the atoms of the material from which it comes. It is literally Einstein's equation of $E=MC^2$. (See lines 19-26 of page 29 of applicant's specification.]

It is obvious throughout applicant's specification that he teaches and instructs in how to achieve a startling difference in kind rather than one of mere degree, and that his teachings are impressively reinforced by the extensive teachings of the Disclosure Document which is part of his patent application.

The total proof, however, is easily seen in his demonstrations, in that they do EXACTLY as he teaches and instructs in his specification, described above, as an example, in the generator and miniature motor demonstration given me. After, applicant showed me the inner workings of his 700-pound motor coil 305 and 90-pound magnet 300 (which is covered and secured together with fiberglass) and its associated generator coil 306 of approximately 200 pounds. Demonstrations were given as to its operability as follows:

"Evidence before the Patent and Trademark Office and this court IS OVERWHELMING that Newman has built and tested a prototype of his invention in which the output energy exceeds the external input energy; there is NO contradictory factual evidence."

Those words of the Special Master speak for themselves.

The electrical energy output is measured at more than four times the electrical energy input. [Note: This does <u>not</u> violate the Law of Conservation of Matter/Energy if one considers the source of the additional output to be the conductor coil in accordance with $E = MC^2$.]

2) Acting as a motor, Joseph Newman's device performed mechanical work in excess of ten times the electrical energy input.

3) Joseph Newman's device delivers over ten times the torque of a commercial D.C. permanent magnet motor rated at 80% efficiency. However, during this test, Joseph Newman's device is consuming only a fraction of the input power of the commercial motor.

4) These results must be taken seriously. Joseph Newman has made the observation that huge magnetic fields may be generated with minimal power input in a large coil wound with large diameter wire. This coil creates a very large torque on a suitably large permanent magnet. In operation, the batteries powering the coil consume little power and discharge at a very slow rate. Yet, the motor delivers considerable mechanical and/or electrically-generated power.

I shall now proceed to constructively refute the negative doctrines that are a result of the present "Three Laws of Thermodynamics."

A. FACTS:

1. The Three Laws of Thermodynamics were conceived without an understanding of the relationship between heat (gyroscopic massergies/electromagnetic energy) and matter.
2. The Three Laws of Thermodynamics were conceived without an understanding that there is an energy relationship other than the simplicity of [Work = Force X Distance], [Power = Work/Time], and [Force = Mass X Acceleration].
3. The Three Laws of Thermodynamics were originally conceived without any knowledge, understanding, or anticipation of Einstein's equation of $E=MC^2$.
4. The Three Laws of Thermodynamics were originally conceived without an understanding of gravity, electricity, magnetism, inertia, matter, and planetary motion.

QUESTION: If none of these things were understood at the time that the Three Laws of Thermodynamics were conceived, how can these three laws be so "all encompassing" as to be capable of predicting --- on a seemingly "infallible" basis --- the "Doom of the Universe" and the "Total Impossibility of Perpetual Motion?" Those who made such predictions must have understood the mechanical workings of the entire universe.

QUESTION: Did they?

The "First Law of Thermodynamics" (1850) states:

"Energy can be exchanged in the form of heat or of mechanical work, but its total quantity remains constant."

The First Law of Thermodynamics is one of the most positive scientific statements ever made, although this was not the initial intent of this law.

QUESTION: What does this law say?

ANSWER: If one cannot destroy energy, this means that energy always exists. If energy always exists, one can always use it. The facts have indicated to me that the gyroscopic particle composition of all matter is totally in accord with the First Law of Thermodynamics since it appears that the energy (spin speed) of the gyroscopic particle cannot be consumed.

The "Second Law of Thermodynamics" (1850):

The First Law of Thermodynamics proves that the implications of the Second Law of Thermodynamics are incorrect!

The Second Law of Thermodynamics represents a conclusion concerning the use of heat based upon primitive, 19th century mechanical devices. The "Second Law of Thermodynamics" may well apply to such primitive mechanical devices, but it has absolutely nothing to do with the implications of $E=MC^2$.

As I have demonstrated earlier, many of the 19th century scientists believed heat to be only the result of motion. They did not understand that heat was simply the conversion of matter into gyroscopic massergies or electromagnetic energy (heat), as implied by the brilliant work of Joseph Black. Nor did they realize that heat (consisting of gyroscopic massergies or electromagnetic energy) was convertible into matter. They were completely ignorant concerning $E = MC^2$. In their ignorance, they would have said that anyone claiming such a statement was stupid. In my opinion, Joseph Black would have readily accepted the implications of E = MC2.

In 1824, Sadi Carnot published a paper entitled, *"Reflections on the Motive Power of Heat."* Carnot had discovered that heat must flow "downhill," i.e., heat must change from high to low temperatures to perform work. Such a conclusion was based upon the observation of primitive inventions and has no real connection with the essential nature of heat or $E=MC^2$. Joseph Black understood the nature of heat as early as 1760 --- others did not.

By 1850, it was concluded throughout the scientific community that Carnot's discovery of a definite direction for heat flow laid the foundations for one of the basic laws of physics: the Second Law of Thermodynamics. The law was first formulated in 1850 by the German physicist

Rudolf Clausius, who stated, "It is impossible for a self-acting machine, unaided by any external agency, to convey heat from one body to another at a higher temperature."

The essence of the Second Law of Thermodynamics is this: heat will not flow of its own accord from a cold place to a hot one. Again, I repeat that this statement has absolutely nothing to do with the essence of heat and demonstrates a total lack of understanding that heat is gyroscopic massergies (electromagnetic energy), which comprises all matter and that $E = MC^2$.

In physics, it is presently believed that this unidirectional flow of heat, as stated by the Second Law of Thermodynamics, implies the "Doom (or heat death) of the Universe." I vigorously disagree with this unfounded statement! All of the facts now presented in science prove this closed-minded statement to be totally incorrect! This negative statement has been an extreme hindrance to the diligent progress of science since it closes one's mind to creative thought and has succeeded in unjustly influencing young minds that were taught to accept it.

Electromagnetic energy is *perpetually* changing from energy to matter and from matter to energy. [While I fully realize that the use of the word "perpetual" violates current scientific taboos, I will do so anyway!] The gyroscopic entity I have described in this book perpetually spins and travels at the speed of light in accordance with $E = MC^2$. Even if all physical matter could become exactly the same temperature, the gyroscopic massergy (electromagnetic energy) within matter is still moving at the speed of light. Any matter could still be caused to release its incredible electromagnetic energy (gyroscopic massergy) composition!

A chain reaction could be induced within a mass the size of a planet, thereby causing the mass to release its electromagnetic energy (gyroscopic massergy composition) at a rate as rapid as that of the Sun. The mass would then cause a source of heat greater than its surroundings which were retaining the major portion of their gyroscopic massergies (electromagnetic energy) composition within the physical boundaries of the materials. All heat is gyroscopic massergies (electromagnetic energy). All matter is gyroscopic massergies (electromagnetic energy). All matter can release its gyroscopic massergies in the form of heat, light, electrical current, electromagnetic fields, electromagnetic waves, electromagnetic radiation, or in smaller quantities of its total physical form. However, it makes no difference in what form matter is released, since it is always composed of gyroscopic massergies (electromagnetic energy).

The reverse is also true: all gyroscopic massergies (electromagnetic energy) can be converted into physical matter! Having a basic understanding of the ingenious properties of the gyroscopic massergy (electromagnetic energy) composition of all matter in the Universe, the mathematical law of probability tells me that the probability of the Universe undergoing a "heat death" is zero.

One of Joseph Black's important discoveries was that different substances have different capacities for absorbing or emitting heat (electromagnetic energy)!

EXAMPLE:

If 1 kg. of iron at 80°C is immersed in 1 kg. of water at 40°C, then the equilibrium temperature is found to be 43.7°C. In other words, the same amount of heat (electromagnetic energy) has resulted in a much greater temperature change in the iron than in the water.

The same unfounded statement of the Second Law of Thermodynamics is also used in current physics to have stamped the final label of "FUTILE" on the quest for "perpetual motion." I would agree that "perpetual motion" would be futile as long as one accepts the validity of the Second Law of Thermodynamics as explaining everything in the Universe for all time. However, I challenge such validity. It is easy to recognize that in this sense, the Second Law has operationally been a deliberate attempt to close young minds that would be otherwise willing to question the "finality" of the Second Law of Thermodynamics. I am sure that there are many who read this book who have been so unjustly influenced. Please recognize that the conversion of physical matter to electromagnetic energy (gyroscopic massergies) and from electromagnetic energy (gyroscopic massergies) back to physical matter is "perpetual" throughout the Universe and that this phenomenal energy change can be conceptually understood and technologically harnessed in the immediate future for the incredible benefit of humanity!

The "Third Law of Thermodynamics" (developed 1888-1902):

In 1902, measurements of the heat reaction of various substances were examined, and it was found that the free energies experienced an increasingly *small variation* as the reaction approaches absolute zero.

This line of thought was initiated in 1848 by Lord Kelvin (William Thompson). Knowing that when cooled one degree from 0° to -1°C, a gas loses 1/273 of its pressure, Kelvin reasoned that at -273°C, gas should have no pressure and he called -273°C "absolute zero". Scientists at the time further reasoned that if "cold" is simply the absence of "heat," then there should be a point when there is absolutely no heat. This reasoning demonstrates a complete lack of understanding that heat is actually electromagnetic energy (gyroscopic massergies) which comprises all matter and that $E = MC^2$. [Kelvin's knowledge is valuable, however, in terms of designing my pioneering invention where atom misalignment is important, since heat causes random motion and rapid atom misalignment.]

In accordance with the above concept regarding the absence of heat, the Third Law of Thermodynamics was proposed. It states that every substance known to man undergoes entropy, i.e., a measure of the availability of energy to perform work that approaches zero as the temperature approaches absolute zero (-273°C or -459.69°F).

Einstein's equation of $E = MC^2$ and the work I have accomplished prove that this statement concerning entropy is <u>totally incorrect</u>.

Kelvin's results are explained by my prior discussion that heat (gyroscopic massergies/electromagnetic energy) loss from matter causes the atomic entities to demand a smaller area. This is why gases lose pressure at low temperatures, because they are transforming into a liquid state.

The concept that cold is the absence of heat should be corrected as follows: Cold is simply a condition of less gyroscopic massergies or electromagnetic energy (heat) in matter. As long as one has matter, one still has gyroscopic massergies (electromagnetic energy or potential heat). Matter at -459.69°F STILL contains tremendous electromagnetic energy (or heat if properly released) or vast quantities of gyroscopic massergies spinning at the speed of light. Only when matter is gone is when all potential heat gone. The mechanical essence of $E=MC^2$ is the gyroscopic-action-massergy, which is the basic building entity of all matter.

It is totally amazing to me that these three laws of thermodynamics have been so widely accepted, knowing that their total premise is one of negativism, which completely stops the creative thinking processes of a student who is motivated to question or discover a method for a better energy invention that would ultimately be of service to humanity. However, in spite of the negative intentions of those who developed it, THE FIRST LAW OF THERMODYNAMICS PROVES JUST THE OPPOSITE!

It is a positive, scientific statement.

Although this may seem superficially paradoxical, I will make the positive statement that:

> *"There is NO PLACE in science that negativism should be allowed to exist!" The entire history of science has proven repeatedly that whenever it has been thought that something was not possible, it later turns out to be possible. Therefore, as the facts have proven, science should put forth positive statements of hopes and dreams that will perpetually stimulate the creative processes of the human mind. In contrast, throughout my sincere, scientific efforts of nearly two decades, I have had to fight against many negative "scientific statements" that were and are wrong. Such injustice has not been unique to my efforts, but, on the contrary, it has been the common fate of most creative individuals throughout the history of science......"*

Joseph Newman

Building a Magnetic Motor to run an Air Cycle Machine

The costs to develop the software and computer components for the second stage magnetic motor should be around 150 thousand dollars, and all units after that would cost around 2 thousand each to build.

The first stage would be a prototype that can be used to power the air cycle machine - this unit would not be of composite materials and would lack all of the computerized sophistication.

The reason being, is - producing the metal molds for the injection mold process of the thermoplastics would cost around 20 - 60 thousand (this is the price to outsource this to this company: http://epsilonindustries.com) unless you buy the computerized machines yourself to do this which would be a couple hundred thousand at minimum. However, once you have molds for the parts, the price to reproduce a unit dramatically reduces to under 1 thousand each for the composite frame. Then the computer components for the unit would be pricey due to having to have a programmer custom program a system for the network stack and software that runs on each generator - for this, you could either hire a few in house programmers or outsource. The estimate cost of this to be around 300 - 400 thousand - this would be a one-time initial development cost. To reproduce the computer components after this would be minimal, possibly around $400 to computerize each unit. There would also be other development costs for specialized cabling and things of that nature.

It is best to have a less advance unit that you can demonstrate to get more investors, then move onto the much more advance pricey second stage unit.

For the less advance first stage unit, the estimate for materials should be around $13,900. $4400 for 16 of the neodymium magnets http://www.unitednuclear.com/magnets.htm), $7000 for the aluminum blocks, $600 for the copper wire, $400 for a resistance electrical bypass, $1000 for the bearings and other fittings, and $500 for a blower to keep cool air moving over the unit. Then batteries to start the unit - the price for these assuming you are using silicon batteries. The other area of pricing are tool costs, but if you need to buy tools, you can get quotes for those.

The costs to develop the software and computer components for the second stage unit should be significantly less if the programmers are in house, then the development cost for the software will just be equal to their salary. You would need 2 -3 programmers. To initially develop the entire second stage unit the estimated cost would be around 150 thousand dollars, and all units after that would cost around 2 thousand each to build.

This CG model is actually intended to be used in a briefing where you can disassemble, explore, and rotate the unit in real time to aid in explaining how it works. Offset the drum that holds the stator magnets several inches so that you can see the recesses where the neodymium magnets go. The neodymium iron boron magnets use a ferrite sleeve and specific proportions and angles to the magnets to control dispersion of the magnetic field. This is tricky because the magnetic field mushrooms out at circular 90 degree angles and you need to control these fields or else the interference can be great enough that the motor will not work at all.

There is a relatively good solution to solve the coercivity problems with the magnet eventually going into disarray by placing the armature magnets at an every other interval to allow for the magnets a little rest time. This will prevent the magnets from being under constant stress and demagnetizing eventually.

Chapter 7

Rear Wheel Pulley Systems

We want the alternator working with the air bearing pump/compressor when it is being utilized. On the final drive, we want a secondary alternator. They will both maintain battery output, and it can be any place. The second pulley system may not be necessary. The primary alternator will spin with the air bearing pump/compressor. The magnetic motor will spin both the air bearing pump compressor and the primary alternator. The secondary alternator can be spun from the final drive or have its own air drive. The alternator can have its own air bearings. Please understand that the alternator on the air bearing pump/compressor must be standard, unless we can provide an external air pressure source.

Air bearings may be added for creating a zero-friction pulley, a common application for **creating zero-friction pulleys** for counter balances or Web handling.

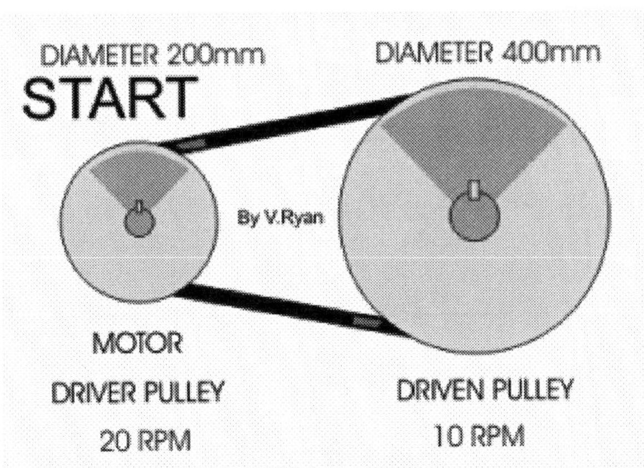

Pulley Systems - Velocity Ratio - 1

The opposite diagram shows a small driver pulley pulling a larger driven pulley. The rpm (revolutions per minute) of the larger driven pulley wheel will be less than the smaller driver pulley wheel. The same principle regarding speed of rotation regarding gears applies to pulley systems as well.

Working Out The Velocity Ratio Of Pulleys

The system shown above has a driver pulley attached to a motor. When the motor is switched on, the driver pulley revolves at 20 rpm. The diameter of the driver pulley wheel is 200mm and the driven pulley wheel is 400mm. This means that for every single revolution of the larger driven pulley wheel, the smaller driver pulley wheel rotates twice. This is due to the velocity ratio. The ratio can be worked out mathematically in different ways.

Chapter 8

Back-up Air System

We need a 12v powered compressor and a ten-gallon air tank. The sole purpose of this system is to maintain our air bearing pressure. This is completely separate from our air bearing compressor. There isn't a more effective method of protecting the air bearing pressure. This engine will rely solely on our air bearings. We only need a total of three ten-gallon air tanks. We only need a ¼" of air line to maintain the air bearing pressure. Location of the tanks is optional and up for examination. It takes a good 5 minutes to drain a ten-gallon air tank through a 3/8" outer diameter air line. Big tanks are not necessary.

CHAPTER 9

THE HISTORY OF COMPRESSED AIR VEHICLES

It cannot precisely be claimed that compressed air as an energy and locomotion vector is recent technology. In fact, at the end of the 19th century, the first approximations to what could one day become a compressed air- driven vehicle already existed, through the arrival of the first pneumatic locomotives. Yet, even two centuries before Dennis Papin apparently came up with the idea of using compressed air (Royal Society London, 1687).

The first recorded compressed air vehicle in France was built by the Frenchmen Andraud and Tessie of Motay in 1838. A car ran on a test track at Chaillot on July 9th, 1840, and worked well, but the idea was not pursued further.

In 1872, the Mekarski air engine was used for street transit, consisting of a single-stage engine. It represented an extremely important advance in terms of pneumatic engines, due to its forward thinking use of thermodynamics, ensuring that the air was heated by passing it through tanks of boiling water, and also increasing its range between fill-ups. Numerous locomotives were manufactured and a number of regular lines were even opened (the first in Nantes in 1879).

Mékarski system tram networks were also built in other towns in France; Vichy (1895), Aix-les-Bains (1896), La Rochelle (1899), and Saint-Quentin (1901).

The H.K. Porter Company in Pittsburgh sold hundreds of these locomotives to coal-mining companies in the eastern U.S. With the hopeful days of air-powered street transit over, the compressed air locomotive became a standard fixture in coal mines around the world because it created no heat or spark, and was, therefore, invaluable in gassy mines where explosions were always a danger with electric or gas engines.

Pneumatic Locomotive

Also in 1896, Porter supplied ten compressed air motor cars for the Eckington System in Washington, D.C. There was a tank on the front of the engine and it was recharged at the station.

Between 1890 and 1902, ten compressed air trams circulated in Bern, Switzerland.

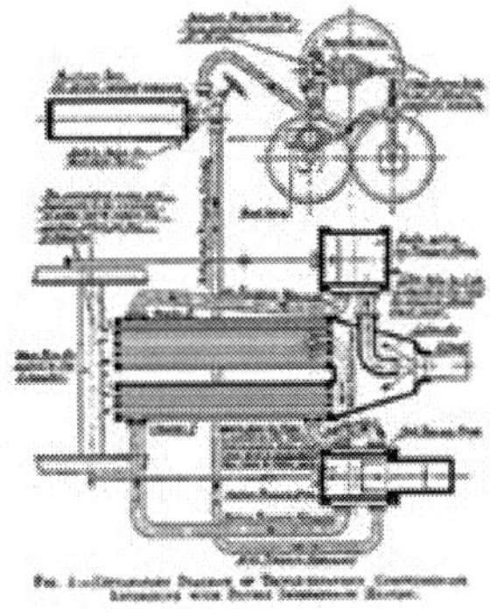

In 1892, Robert Hardie introduced a new method of heating, which, at the same time, served to increase the range of the engine.

However, the first urban transport locomotive was not introduced until 1898, by Hoadley and Knight, and was based on the principle that the longer the air is kept in the engine, the more heat it absorbs and the greater its range. As a result, they introduced a two-stage engine.

Charles B. Hodges will always be remembered as the true father of the compressed air concept applied to cars, being the first person not only to invent a car driven by a compressed air engine, but also to have considerable commercial success with it.

Later on, in 1912, the American's method was improved by Europeans, adding a further expansion stage to the engine (3 stages).

After years of working on a system for driving an automobile by means of compressed air, Louis C. Kiser, a 77- year-old from Decatur USA had succeeded in converting his gasoline engine into an air compressed system. Kiser removed the entire gasoline line, the cylinder head, water-cooling system, and self starter. A special cylinder head is substituted and a compressed air tank added in place of the gasoline tank.

In 1926, Lee Barton Williams of Pittsburg USA presented his invention: an automobile which, he claimed, runs on air. The motor started on gasoline, but after it had reached a speed of ten mph, the gasoline supply shut off and the air started to work. During the first test, his invention attained a speed of 62 mph.

The first hybrid diesel and compressed air locomotive appeared in Germany in 1930. The pressures brought to bear by the oil industry in the transport sector were ever greater and the truth of the matter is that they managed to block investigation in this field.

Compressed AIR MOTOR Runs Car

In January 1932, what appeared to be the first journalistic article ever written about a car driven by compressed air was published. In 1932, J.M. Custer of Piggott, Ark. had an idea for a car that would run on compressed air. He not only built the car, but it actually ran. He used an engine that was the result of six years of research by Roy J. Meyers, which resembled a radial airplane motor in appearance. It was mounted upright in the same space occupied by gasoline motors in standard cars.

Four air tanks filled with compressed air powered the car 500 miles at a speed of 35 miles an hour. The engine did not require a cooling system, ignition system, carburetor, nor the hundreds of moving parts included in a standard gasoline motor. The compressed air took care of all of those features and left a vehicle that cost nearly nothing to maintain or use.

That year, they conducted an amazing demonstration in Los Angeles using a standard automobile chassis that was powered with a newly developed compressed air motor. It whizzed around the city streets using nothing but the compressed air and requiring no gasoline even though fuel prices back then were dirt cheap. The motor headed the air tanks until a pressure of 200 PSI was obtained. As the warm air went through the engine and cooled, it was recovered and drawn into a compressing chamber, where it was reheated again and returned to the tank.

That technology, which didn't attract much interest 75 years ago is resurfacing using today's high technology by a research and development company called Magnetic Air Cars, Inc.

In 1934, 21-year-old Johannes Wardenier announced that he developed the world's first fuel-less automobile. For weeks, Dutch newspapers reported of an incredible invention that would change the world forever. Not long after, he was mysteriously imprisoned in a mental institution, his design for the engine was stolen and he was kept under constant guard and never allowed to see anyone.

Later, he was sent off to a concentration camp where he remained until he was near death and his idea for a motor that ran on air was long forgotten.

According to his design, the hot air was pressed into a motor which contained a number of cylinders, half of which go down when the others rise up. As in an ordinary engine, the crankshaft forced a rotating movement. The major difference was that the air, after having passed through the cylinders, passed again through the cylinders by means of a compressor at the side, causing a continuous circulation and enough perpetual movement to last three months.

After the Second World War, the term "air engine" was never again used in textbooks referring to compressed air or pneumatic locomotives, and whenever they were mentioned, the article would go on to state that these engines were of little use or efficiency.

In the 1970's, Joseph P. Troyan designed an air-powered flywheel that could propel an automobile using the principle of "ratio amplification of motion in a closed system". The Troyan air mobile (U.S. Patent No. 040011) was easily attached to electrical generators to create a pollution-free, variable-power system. "My engine simply uses nature's most powerful force: air pressure", Troyan commented.

In the 1970's, Willard Truitt presented his invention in McKees Rocks, USA. However, because he did not have the financial means to develop his compressed air car further, he gave the rights of his invention to NASA and the US Army in 1982.

After noticing steam pumps in Texas refineries, Russel R. Brown started thinking about the idea of an air-powered engine. Twenty years later, in 1974, his invention was finished. Brown claimed that his engine could not only run on air, but could also be its own source of power. "We're plagued with pollution, but my engine does no more damage to the air than if you inflate a tire and let it out again¨, Brown comments.

In January 1975, driving on compressed air was proposed by Sorgato in Italy as a viable fuel-economy alternative to the electric car for industrial and urban use. The first experimental model had nine air bottles charged to 2840 psi by an external compressor. The maximum speed of this near-silent and non-polluting vehicle was said to be 30 mph and had a duration of around two hours.

In 1976, Ray Starbard from Vacaville, California developed a truck that was able to drive on compressed air. He felt that he had invented the power system of the future, a system that would greatly change the automotive face of the world. ′It's the car of the future, there's absolutely no doubt in my mind′, Starbard comments, ′and all because of Mother Nature's purest gift – air¨.

In 1979, Terry Miller decided that compressed air was the perfect medium for storing energy. He developed Air Car One, which he built for $1,500. Terry's engines showed that it was feasible to manufacture a car that could run on compressed air. He patented his method in 1983 (US4370857).

In the 1980's, Carl Leissler developed a motor that was able to function on air. The retired horticulturalist had been working from his garage in Hollywood for over 15 years. He says that to use his motor in a car, you might have to use a small electric or gas energy source to help drive the air compressor. 'We might be able to get 2000 miles per gallon, air is a power in itself', Leissler comments.

Oblivious to curious stares, Claud Mead drove his compressed air car through the streets. The aim of this car was to spare the American motorists from gasoline woes. The big version of the car would be able to go 800 miles on a full tank. Australian inventor Des Hill had been working on a compressed air motor for thirty years and had spent more than $20,000 on his project. The air compressor, which worked when the engine was running, would ensure that the cylinder was filled to capacity at all times. Thus, Hill's engine would realize the principle of perpetual motion.

George Miller said the air car that he invented in 1984 would run forever. Miller envisioned his invention to be a solution to fuel and pollution problems, and the end to wars in the Middle East. The 58-year-old retired coal miner and bricklayer from Johnstown, USA used a medium-sized, four cylinder engine out of a twelve-year-old Opel. Air from one of the tanks is fed into it through the spark plug holes, and air pressure moves the piston. The air is circulated out of the engine and back into the tank.

Jet aircraft starter motors using compressed air as a fuel provided unique power for this experimental dragster from Phoenix, US that had covered the quarter-mile distance in the nine second range.

In May 1987, an article was published about Miami inventor Ricardo Perez-Pomar. The 61-year-old pneumatic engineer from Cuba claimed to have developed an engine that will continuously refill the very tank of compressed air that powers it.

'This machine can be in full operation for months and months before its air tank must be artificially refilled', Perez promises.

Until 1987, the German company Arnold Jung Lokomotivenfabrik GmbH produced locomotives functioning on compressed air to be used in mines. In the 1980's, they were still selling and renovating locomotives.

Currently, the tram association in Bern, Switzerland (BTG) is developing a locomotive according to the original plans. It is expected to be ready in 2010.

At present (2008), various persons and companies are developing compressed air motors applicable to transportation, apart from the many companies that produce and commercialize compressed air motors for industrial purposes.

CHAPTER 10

AIR BEARING PUMP/COMPRESSORS

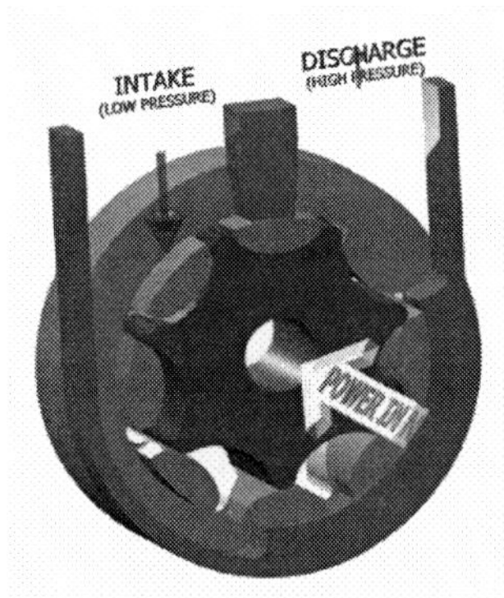

The air bearing compressor needs to be able to fill at least two ten-gallon tanks to 200 psi. We are concerned about having an output speed and an output potential. We are trying to achieve 200 Horse Pressure™ that maintains 65 mph. Everything else is gravy. With this solution, we do not have to put up with connecting rods, pistons, and displacement, timing or thermal dynamic efficiency because they do not apply. We can use off-the-shelf parts to achieve an idle. The idle will maintain an automatic transmission and keep it engaged. If it puts out 1000 cfm and 200 psi to fill the tanks, then it should be capable of maintaining the idle at all times.

Once we request maximum output from the engine (stab the throttle), several events occur:

The valves are opened at the 2 ten-gallon tanks.200 psi is sent to the drives of the air amplifiers. There are two on the primary and one on the secondary. The orientation of the duel primary air amplifiers are negative to positive. They manifold together and feed the secondary air amplifier. This pressure is captured behind a priority valve on the primary intact of the entire system.

***Priority Valve.**

The priority valve remains closed until the engine changes pressure to vacuum. The vacuum lets the priority valve swing open. The engine now draws air as it needs. The priority valve inlet is a 3″ tube. The end of the inlet is a 40 degree bologna cut. The valve chamber is a 3.5″ square. The opening has a silicon seal. The valve door rests against the seal. When you have pressure on one side, it puts pressure against the seal and seals tighter. As the engine begins to draw, it changes pressure to vacuum. The priority valve doors swings open. The engine now draws air on its own. As a result, the star rotor is no longer needed to motivate the vehicle. Its discharges close. The pressure is now used to recharge to 2 ten-gallon tanks, and therefore, it revitalizes (recharges) itself. The throttle is achieved by governing the air pressure to air amplifier drives.

Cruise Mode.

The priority valve is open. The engine draws air as it needs. The exhaust shares the same chamber as the intake port. Therefore, they share the same air cleaner.

Air filter and muffler.

Essentially the same system. Soft noise coming out the back.

Cabin Environment.

A ½" hose can feed the cabin and provide fresh air. This is an amazing green technology for cleaning the air. Our air bearing compressor can process any vapor or gas with the only associated design consideration being selection of a compatible material. It works by using inner and outer rotors that rotate about their respective axes. A drive mechanism synchronizes the rotors so that they do not touch. Abradable coatings form the seals between rotors and stationary porting components. As they rotate, the rotors create a void that opens, reaches a maximum volume, and then closes. Inlet gas enters through the intake port as the void opens. Once the gas is captured, the void volume decreases, causing the pressure to increase. After the design pressure is achieved, the gas exits through the discharge port. Notice that a void opens eight times per rotation of the outer rotor, which allows the compressor to process large volumes of gas.

The position of the leading edge of the discharge port determines the compression ratio. If the leading edge is positioned to make the discharge port large, the compression ratio is small. If the leading edge is positioned to make the discharge port small, the compression ratio is large. By using a sliding mechanism, the leading edge position can be changed on the fly, giving the compressor a variable compression ratio. Besides being shaft driven, we are researching the possibility of using an integrated motor to drive the compressor.

Benefits:

1. The air bearing compressor is **efficient**
2. It can handle a **high compression ratio** (about 6 to 1 in a single stage).
3. It is **compact**.
4. It is **inexpensive** to manufacture.
5. It has a **high turn-down ratio**.
6. There are **no valves** to limit the gas flow.
7. The compressor is **scalable**.
8. It **does not require liquid injection** to achieve the desired output. (Liquid injection improves efficiency, but is not required.)
9. **Pressure and flow are independent**.

Flexible - materials are selected to be compatible with a wide range of fluids and temperatures.

Chapter 11

Subsystem 2

We could have a nozzle like the AirTX air amplifier blowing on the power drive of the air amplifiers, which could blow through the secondary discharge of the air bearing compressor, which will be entrained around it. We have an air bearing and a thrust air bearing. We have a flat surface on the vein of the turbocharger. On the other side, add an extra 3/8" of thickness, approximately 55% the diameter of the turbocharger of the impeller and the turbine. Our modification tube, the turbine and impeller, and our drive system fit on the inside of the impeller and the turbine. It fits on all of them. The tube, turbine and impeller combine for a total of six drives, since we are using two of each which are the positive and the negative.

With the old method, they did not have direct captured static pressure. They were only trying to capture flow. You can only capture pressure, and not flow.

As soon as you capture the pressure and entrain it, you have static pressure. The previous design could not achieve that. Our system has the ability to self-start, as well as the ability to self-maintain.

We have two air flow amplifiers forming the substation 2. Substation 3 consists of a single air flow amplifier that should provide our projected output. If we need more kinetic energy, we can stage up the entire process. We can do it both through size, chamber size, and size of amplifiers.

Note: The air flow amplifiers consist of a combination of an air flow amplifier and air bearing turbocharger. Initially, we will be using 2 for stage 2, and 1 for stage 3. The combination of the entire stage 2 and stage 3 is called our air cycle machine motor. Subsystem 1 must continue to provide high compressed ambient air to Subsystem 2. The compressed air is delivered to Substation 2. Subsystem 2 consists of an air flow amplifier that entrains the compressed air and expands its volume significantly. The air flow amplifier may run at speeds exceeding Mach 1 (760 mph) and deliver a payload of 500 psi.

This Center Stage Subsystem 2 is in the center of Figure 1. Our claim is that once started, this Subsystem 2 produces enough amplified compressed air to power Subsystem 3.

Could a car run on compressed air?

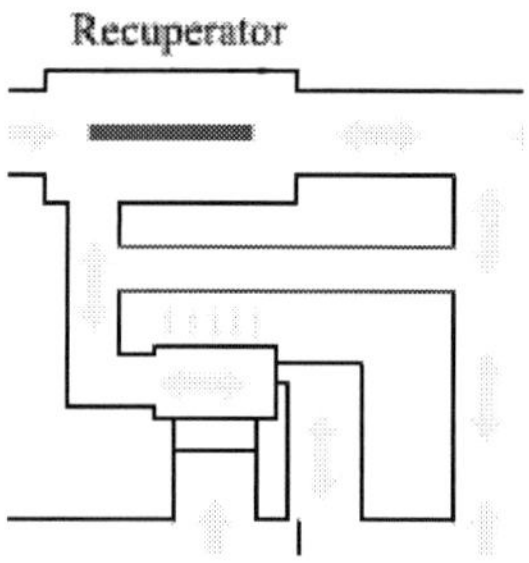

Yes, it could. You could compress the air at your house using an air compressor, fill a compressed air tank in the car, and the car could run off of it. You could use an engine very similar to a steam engine (using pressurized air instead of pressurized steam) to convert the compressed air to rotational energy.

The main problem you would have is the size of the "air tank" you would need to get any kind of reasonable range. You would need a lot of compressed air to take a car very far.

We believe that we may have found a solution to this problem. An air compressor dedicates only 20% of compressed air to rotational energy. 80% of the air is lost and/or wasted. Through our air cycle machine, we may recapture the lost 80% and recycle it back through the air compressor as cold air for a continuous loop of operation. The cold air also allows us to manufacture our devices with strong, lightweight, and cost effective material.

The 20% of the used air is fed through a frictionless air bearing fan blade. The shaft comes off of the fan blade and connects into the tranny. The frictionless air bearings in the tranny give the vehicle its power.

This method eliminates the need for a combustion motor and runs the air compressor and air flow device as a direct drive to the tranny. This method also keeps the shaft and harmonic balance moving the fluid through the tranny. The air compressor will never run out of air utilizing the Coanda system of operation. This means that a car may run without the need for expensive gasoline, or the need to recharge a battery.

We believe that the velocity and speed of entrained air into the Coanda inlet housing may be 500 psi and may flow through the Coanda inlet housing with no back pressure, breakdown, or distortion. The air flow may reach speeds in excess of 760 mph, or Mach 1 speed.

A compressed air car would generate no pollution while in operation. The magnetic engine that produces the electricity that compresses the air also generates no pollution.

Maintenance is absolutely minimal. There is no oil to change, no radiator to clean and fill, no transmission to foul up, no fuel pump, no water pump, no carburetion problems, no muffler to replace, and no pollutants emitted into the atmosphere.

The best alternative is to combine the magnetic engine with the air flow device. This is the basis of our Pilot Project. We are developing a new type of car that will run on a magnetic engine that generates electricity for compressed air. In this business plan, you will learn about the technology behind magnetic engines and compressed air cars currently being developed and how they may replace your gas guzzler by the end of the decade!

We believe that magnetic engines and air flow devices are simply the greatest advancement in technology in the history of mankind.

Air Flow Amplifier

Air amplification gives you high performance with low noise

The AiRTX™ Air Amplifier releases a tiny amount of compressed air at **near-sonic** velocity through an adjustable, internal, ring-shaped nozzle. The high-speed "tube" of air released through the front creates a strong vacuum behind itself, pulling additional surrounding air through the rear of the amplifier, while pushing the ambient air in front.

The Air Research Technology Company
www.airxinternational.com 1-800-307-7475

IAC Industrial Automation Controls
www.iacohio.com 1-800-644-6215

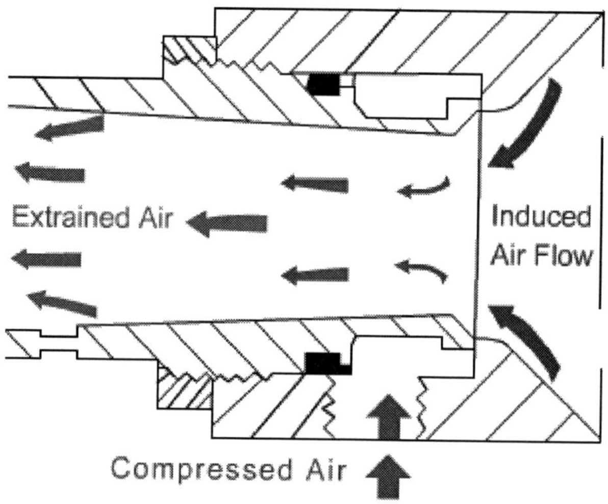

Chapter 12

Subsystem 3

The Final Stage Subsystem 3 is our air bearing turbocharger. This Subsystem 3 is on the right in Figure 1. Its purpose is to amplify the compressed air received from our air flow amplifier in Subsystem 2. Subsystem 3 may amplify Subsystem 2 by a factor of five. Our claim is that only 20% of the compressed air in Subsystem 3 is needed for mechanical shaft power.

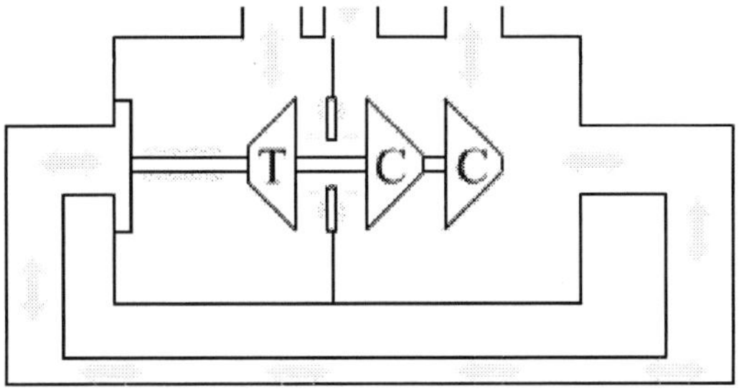

The Final Stage Subsystem 3

We believe that the Starting Stage Subsystem 1 may deliver compressed air as claimed. We believe that it will do so with efficiency greater than a conventionally-driven compressor. A quantitative description of the battery, electric magnetic engine, and compressor enables us to make only an estimate of the overall efficiency of this part.

An efficiency report may be prepared when all three subsystems are operational and the Pilot Project has been completed.

Likewise, the Final Stage Subsystem 3 on the right may deliver power to a load, assuming an adequate input flow of compressed air at suitable pressure and volume is supplied.

What makes the Final Stage Subsystem 3 so unique is in our ability to stage up the air flow amplifier and air bearing turbocharger for the purpose of reclaiming ambient air and recycling the compressed air through a recuperator. 20% of the compressed air is used for mechanical work, while the remaining 80% is reclaimed and recycled.

The purpose of the transmission is to multiply this power to the load at the requested combination of speed and torque.

We believe that the proposed design for these subsystems would accomplish this with efficiency advantage over current technology. A more specific quantitative description of the two elements, the air cycle motor and transmission, would be required to make any estimate of the efficiency of these subsystems. No efficiency report may be quoted by the inventors for these subsystems until after the Pilot Project has been completed.

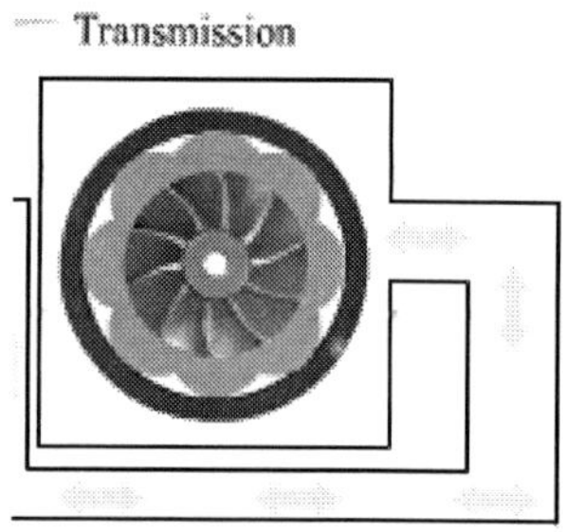

The air cycle machine consists of an air flow amplifier (Subsystem 2) and air bearing turbocharger (Subsystem 3). The air cycle machine is the heart of the system. Our claim is that the air flow amplifier and air bearing turbocharger are individually efficient enough for each to power the other for a distance of at least 200 miles before needing additional compressed air from subsystem 1. Subsystem 1 must provide enough compressed air to the air cycle machine to drive the load at all times.

Our claim is that we may be able to deliver enough compressed air to the air cycle machine to drive the load. Our claim is that the compressor may deliver enough air (via the air flow amplifier) to power the air bearing turbocharger for a significant distance before needing additional compressed air from Subsystem 1. This is what makes our engine far more efficient than what is currently available on the market today.

Our claim is that the amplified compressed air may be recycled and reclaimed through a recuperator. Only 20% of the compressed air is needed to drive the load. 80% may be recycled and reclaimed to drive the load for at least 200 miles before Subsystem 1 is needed once again to feed Subsystem 2 and Subsystem 3 and start the process anew.

It is our claim that the air cycle machine does not violate the first law of thermodynamics, conservation of energy. We believe Subsystem 1 may be capable of delivering energy to the air motor/transmission (Subsystem 2 and 3) in the form of compressed air on a continuous basis.

Outside of nuclear processes, you can never create (or destroy) energy. You can only convert it between different forms, and this conversion can never be done with perfect efficiency, i.e., 100% of input energy can never be converted to the desired output form. A conventional combustion engine, for example, converts the chemical energy of fuel to heat by burning and then converting a portion of this heat energy to mechanical energy by means of the pistons, crankshaft, etc. of the engine. Most of the heat energy is lost to the ambient either through the

exhaust heat (which is required by the second laws of thermodynamics, see below), or through the radiator (which is required by the practical temperature constraints on the engine).

We believe our proposed air cycle machine may run for a significant length of time even before delivering bleed air to the air motor.

The efficiency in the air bearing turbocharger and especially of the Air Flow Amplifier allows the load to run for a significant length of time. The air volume multiplier (Coanda entrained flow amplifier device) would lower the inlet temperature to the Air bearing turbocharger. The Air bearing turbocharger is far more efficient when lower inlet temperature is added. This actually increases the cycle efficiency further.

Combustion Turbine manufacturers strive for the highest possible turbine inlet temperature (and have done so ever since the jet engine was invented in the 1930s) because this contributes directly to cycle efficiency.

We believe our air bearing turbocharger strives for the lowest possible turbine inlet temperature. Our air bearing turbocharger demands the lowest possible turbine inlet temperature for maximum efficiency using compressed air as our motivational source.

A simple interpretation of the first law: you can never win the game, that is, you can never come out with more energy than you went in with. We believe that there is enough energy present in Subsystem 1 to support Subsystem 2 and Subsystem 3.

We believe that all of our Subsystems have the ability to recapture and recycle lost energy more efficiently than today's combustion engines. We believe that we have discovered a unique and novel way to convert and harness this energy to do mechanical work. Compressed air cars are currently running in France and in India. These air cars run 200 miles before needing to refill their air tanks. We believe our solution is far more efficient in converting energy to mechanical work as long as there is enough compressed air available to support the load. We believe that we may be able to provide enough compressed air to support the load.

The simple interpretation of the second law is that you can never break even in the game. Efficiency of a heat or power cycle can never be 100%, let alone greater than 100%.

We believe that our Magnetic Air Car engines do not violate the second law of thermodynamics.

Microprocessors and digital signal processors may be used to add sophisticated computer intelligence to the magnetic motor and the air cycle machine. This may allow the magnetic motor and air cycle machine to capture energy that is normally lost and recycle that energy more efficiently.

Our website may also monitor hardware and update the latest software to the engine to improve efficiency in real-time. We may also integrate a UHF wireless network for two-way audio, video and data with human interface capabilities.

Our Magnetic Air Cars may also be used to clean up the atmosphere through air filter ionization techniques while the car is in motion.

Our cars may also be tracked utilizing GPS for authentication between sender and receiver. Engine may be shut off and disabled through GPS authentication for security mode made valid. GPS also may be able to provide constant telemetry of vehicle while in motion.

CHAPTER 13

AIR BEARINGS

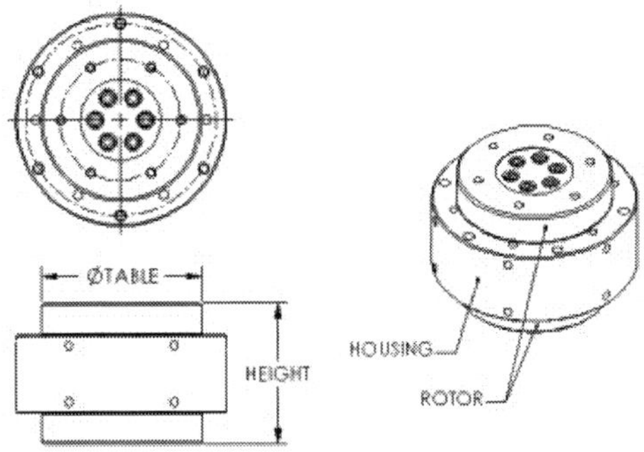

Building off of this preliminary testing model, we can then add air bearings, lightweight parts consisting of a compressor, turbine, and turbine housing, and the air flow device. These rotary air bearings are designed for accuracy, precision and ease of use. They can be used in any orientation and are easy to integrate with motors and encoders for complete positioning solutions.

1. Can air bearings handle the load without catastrophic failure?

We believe the concept behind Magnetic Air Cars magnetic engines and air flow device is feasible. We believe that our concept may be a highly efficient way of producing electric power and operating a transmission. We believe our concept to be non-polluting and environmentally friendly. We believe that we may use an alternating current (AC) generator or a direct current generator (DC). We believe that we may be able to connect multiple units together for larger power output.

Bearing technology represents one of the age-old problems for mechanical engineers. Rolling element bearings developed in the last century were a revolutionary improvement over the plain bearings that had been pushed to their limits in applications like electric motors and automobile wheels. Similarly, rolling element bearings are being pushed to their technical limits even today by the demands of applications like semiconductor manufacturing, high-resolution scanning, and high-speed machinery.

Air bearings represent the next logical step in bearing design. Air bearings in general have a proven track record, having been employed in coordinate measuring machines for 20 years. The many technical advantages of air bearings such as near-zero friction and wear, high speed and high precision capabilities, and no oil lubrication requirements are powerful advantages for today's machine designers. However, these benefits have not been more fully utilized to date because air bearings are difficult to manufacture and have not been commercially available until recently. New Way Precision was founded ten years ago to pioneer the use of porous media technology and make air bearings that are robust, simple to use, inexpensive, and available off-the-shelf. The purpose of this guide is to answer the common questions that designers have when first considering air bearings, as well as to provide detailed information that will help ensure the success of your advanced application. This is believed to be the first such collection of information regarding the selection, mounting, and application of commercially available air bearing products.

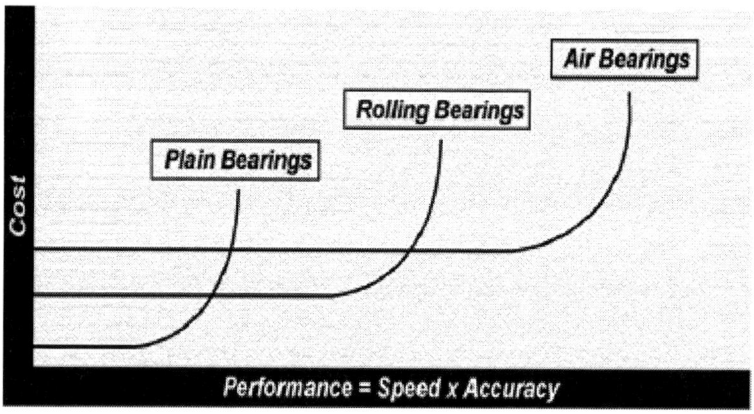

Figure 1 - Bearing Performance vs. Cost

Figure 1 - Bearing Performance vs. Cost

2. *What is an air bearing?*

Unlike contact roller bearings, air bearings utilize a thin film of pressurized air to provide a 'zero friction' load bearing interface between surfaces that would otherwise be in contact with each other.

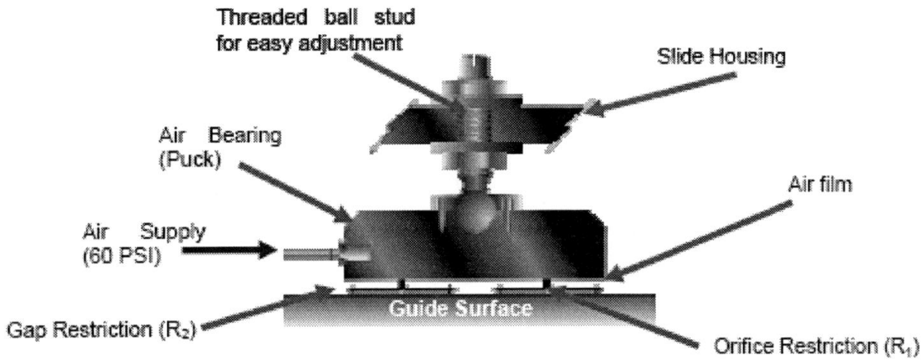

Figure 2 - Flat Air Bearing

Being non-contact, air bearings avoid the traditional bearing-related problems of friction, wear, and lubricant handling, while offering distinct advantages in precision positioning and high speed applications. The fluid film of the bearing is achieved by supplying a flow of air through the bearing face and into the bearing gap. This is typically accomplished through an orifice or a porous media that restricts or meters the flow of air into the gap, referred to in Figure 2 as R1. The restriction is designed in such a way that, despite the air constantly escaping from the bearing gap, the flow of pressurized air through the restriction is sufficient to match the flow through the gap. It is the restriction through the gap (R2) that maintains the pressure under the bearing and supports the working load. If air pressure were introduced to the gap without restriction (R1), the flying height would be higher, the air consumption higher, and the stiffness would be lower than what could be achieved with proper restriction. This restriction is referred to as air bearing compensation.

It is used to optimize the bearing with respect to lift, load, and stiffness for particular applications, and will be discussed in further detail later.

3. Why use air bearings?

Because of their advantages over rolling element bearings, air bearings are a natural choice for applications such as coordinate measuring machines, precision machine tools, semiconductor wafer processing machines, and other clean room-style, high speed, and precision positioning environments. The main advantages of air bearings are listed below. Some of the specific concerns of the design engineer (friction, wear, stiffness, and load capacity) are then discussed in more detail.

ZERO FRICTION....

Because of zero-static friction, infinite resolution and very high repeatability are possible.

ZERO WEAR....

Non-contact means virtually zero wear, resulting in consistent machine performance and low particle generation.

STRAIGHTER MOTION....

Rolling element bearings are directly influenced by surface finish and irregularities on the guide. Being non-contact, air bearings minimize these errors.

SILENT AND SMOOTH OPERATION....

Re-circulating rollers or balls create noise and vibration as hard elements become loaded and unloaded and change direction in return tubes. This is especially noticeable in the resolution of scanners.

HIGHER DAMPING....

Being fluid film bearings, air bearings have a squeeze film damping effect resulting in higher dynamic stiffness and better controllability.

ELIMINATES OIL....

Air bearings do not use oil lubrication, eliminating the problems associated with oil. Industry environments (dry machining) ways are dry and bearings are self-cleaning because positive air pressure pushes dust away. In contrast, oil lubrication becomes a lapping slurry. High speeds =high acceleration. No balls or rollers to slip at high acceleration.

FRICTION....

Variances in friction have always been at the heart of precision positioning problems, particularly when attempting to initiate or stop motion precisely. This is because in roller bearings, and especially in plain bearings, the static coefficient of friction is higher than the dynamic coefficient of friction. In other words, it takes more force to initiate motion than it does to maintain motion. So when a motor turns a screw to push a slide, the screw winds up and stores some of the energy because of the high static friction. When the slide starts to move, the friction falls and the screw unwinds, pushing the slide past its desired position. This phenomenon is known as 'stick-slip' and is most pronounced in plain way systems. However, it can also cause positioning or bandwidth issues, even in machines with rolling element bearings and closed-loop feedback. Today, heavy machine tool slides can be positioned within .0001" using rolling element bearings because the difference between the static and dynamic coefficients of friction is reduced by an order of magnitude compared to plain bearings. Yet, even rolling element bearings are reaching their limits. For instance, in some areas of the electronics capital equipment industry, positioning to even .00001" is considered too coarse. Rolling element bearing manufacturers have started to reduce their preloading compromising stiffness in what has become known as a 'California fit' in an effort to meet these requirements, but there are limits to how effective this can be. In air bearings, there is no difference between static and dynamic coefficients of friction, so the stick-slip issue is completely eliminated. Friction in air bearings is a function of air shear from motion, so at zero velocity, there would be zero friction, making infinite motion resolution theoretically possible. Friction has a direct effect on efficiency. In fact, one of the first air bearing patents was applied for by Westinghouse for use in vertical steam- driven generating turbines. Westinghouse knew that because the viscosity of air is several hundred times lower than that of oil, he could reduce the energy loss due to oil shear. Unfortunately for Westinghouse, at that time (1890), it was exceedingly difficult to manufacture the large bearing surfaces to the high precision required for air bearings to work. Today, large turbines still use oil-based hydrodynamic bearings, but many of the new micro turbines on the market are employing aerodynamic bearing systems to improve efficiency. Friction also has an effect on precision. Remember the old saying, "What are the three main sources of error in a precision machine? Heat, heat, and heat!" Friction creates heat, which is a precision engineer's worst enemy. For instance, when a spindle heats up, it grows axially. As the heat conducts into the headstock, it will expand and the center of rotation will grow away from the base. In most cases, air bearings will create significantly less heat in a given application than a rolling element or plain bearing. This is not to say that air bearings cannot create heat. They can, but relative speeds need to exceed 100 feet per second before significant heat can be measured.

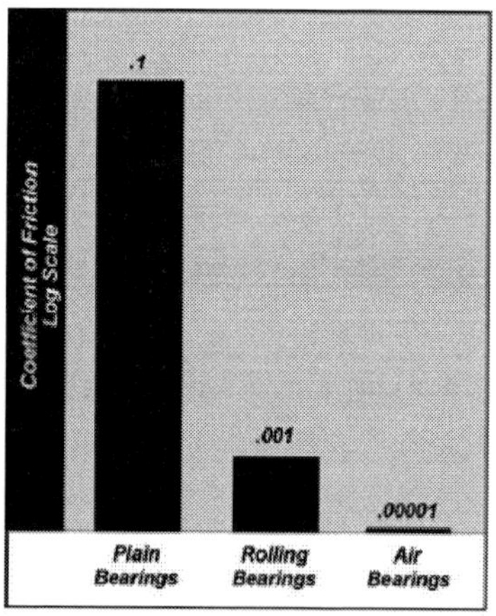

Figure 3 - Coefficients of Friction

WEAR....

Mechanical wear is another thorn in the side of the design engineer. Advanced machines are requiring faster speeds and higher reliability. In fact, there are machines currently on the market that make a billion moves per year. It is impractical to perform accelerated testing on such a machine, so at best, the engineer must employ speed, acceleration, and loading calculations in order to estimate the life of the bearing. This problem is avoided with air bearings due to their non-contact nature. Speed, acceleration, and loading are not wear factors and they have no influence on the life of the air bearing. The mode of wear in an air bearing is erosion, so the cleanliness of the air has the greatest effect. Air bearings are immune to conventional notions of wear and will perform exactly the same in the 10th year of operation as they did in the first, even after a billion cycles per year. This is a big advantage when it comes to machine reliability. Equipment builders that can claim to their customers that wear has been eliminated as a variable that could affect their statistical process control will have a distinct competitive advantage. The lack of debris from wear and the lack of need for oil lubrication due to the noncontact nature of air bearings also mean that they are ideally suited for use in clean rooms, medical, pharmaceutical, and food processing environments. Air bearings also excel in dry, dusty environments such as salt or sugar factories, which can be highly corrosive. In these environments, any oil lubrication quickly becomes a lapping slurry. Air bearings have a self-purging effect with constant air exiting the bearing, which blows away light, dry dust.

STIFFNESS....

A common misconception about air bearings is that they do not have the required stiffness for precision applications. However, a 6" diameter bearing running at 60 psi actually has a stiffness greater than 2,000,000 lb./in. with a load of 1,000 lbs. Put another way, that's less than one-half of a millionth of an inch of deflection per pound of additional force. By testing a bearing through a range of loads and measuring the change in air gap, a stiffness plot can becreated like the one shown in Figure

4. Notice that the slope is not linear and as the film gets thinner, the stiffness gets higher. Pressure and surface area both affect stiffness proportionately, but the most significant factor in air bearing stiffness is the idea and use of compensation. Compensation is a way of controlling the flow of air into the air gap and is the true key to air bearing stiffness. The object of compensation is to create a restriction of the airflow into the gap before the restriction of the gap itself. The air gap must be a restriction, otherwise the pressure would not remain under the bearing and instead, would equalize with the ambient pressure. But how can the pressure reserve created by a restriction of airflow into the air gap provide stiffness? Consider the case where the flow through the orifice at 60 psi is equal to the flow through the gap with a 150-lb. load on a bearing flying at 300 micro inches. A 2.5"- diameter bearing has nearly 5 square inches of surface area so the average pressure under the face of the bearing will be 30 psi. Now imagine that the load increases to 200 lbs. This increases the average pressure under the air bearing to 40 psi and compresses the air gap to 200 micro inches. The reduced gap creates increased restriction and reduces the flow. The reserve pressure that had been held back by the orifice now allows for increased pressure in the gap, creating a restoring force that gives the air bearing stiffness.

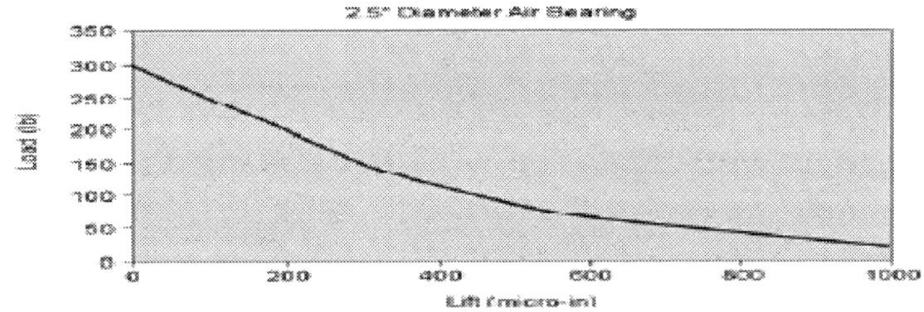

Lift load curves are useful for sizing air bearing. Notice also that the slope of the curve is representative of bearing stiffness. A vertical line would represent infinite stiffness and a horizontal line would represent zero stiffness.

$$\text{Stiffness} = \frac{\Delta \text{ Load}}{\Delta \text{ Lift}}$$

Load (lbs)	Lift (micro-in)	Δ Load	Δ Lift	Stiffness (lbs/in)	Compliance (micro-in/lbs)
100	450				
		25	90	277,778	3.6
125	360				
		25	60	416,667	2.4
150	300				
		25	50	500,000	2.0
175	250				
		25	50	500,000	2.0
200	200				

The stiffness between any two points on the curve in figure 2 can be calculated as shown. Stiffness is typically expressed in pounds per inch units, but it is more intuitively obvious when described in micro inch per pound terms. So rather than saying a bearing has 500,000 lbs/in stiffness we may say it will displace two micro inches for each pound load.

Figure 4 - Air Bearing Stiffness

LOAD CAPACITY....

It can be seen from the above diagram that compensation compromises load capacity for the benefit of stiffness and stability. Even without compensation, the maximum theoretical load capacity of the bearing could not be described as surface area times input pressure. A load equal to this value would ground the bearing. The air bearings will not actually carry the full theoretical load because there is leakage around the edges of the bearing that prevents the pressure from being equal across the whole face. As the air issues from the orifice, it expands through the gap, traveling away from the orifice and giving rise to pressure gradients in the gap that can be mapped as pressure profiles (figure 6). These pressure profiles are influenced by the amount and location of orifices and grooves, and especially by the type of compensation used, with porous compensation giving the most consistent pressure distribution under the bearing. When compensation is used to increase bearing stiffness, the average pressure under the bearing is usually about 50% of the input supply pressure. Therefore, air bearings can generally be considered to be 50% efficient with regard to input pressure. This rule of thumb is modified by the shape and size of the bearing. Bearings that have a high percentage of their area near the edge of the bearing, such as small or narrow bearings, will have lower efficiency, while larger bearings will have higher efficiencies. This is one of the reasons that it can be more challenging to make smaller rather than larger bearings. Although the load capacity of air bearings is limited compared to rolling element bearings, they carry the same load per unit area as traditional plain bearings for machine tools. This is usually more than sufficient for today's high-speed, lightweight machine applications.

Load Capacity

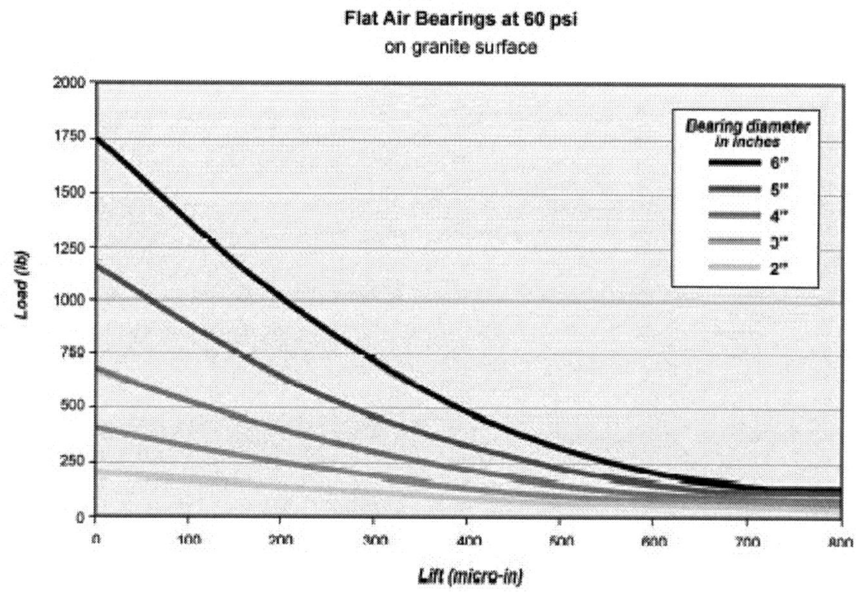

Figure 5 - Lift-Load Curves

Surface area x input pressure = grounding force
Surface area x input pressure x efficiency = load capacity

4. Types of air bearing technology

There are two basic types of precision air bearings: aerodynamic bearings and aerostatic bearings.

Aerodynamic bearings

Aerodynamic bearings depend on relative motion between the bearing surfaces and usually some type of spiral groove to draw the air between the bearing lands. This bearing action is very similar to an automobile hydroplaning on a puddle of water at high speed. At a lower speed, your tire would cut through the water to the road. In just this way, aerodynamic bearings require relative motion between the surfaces, when there is no motion or when the motion is not fast enough to generate the air film which the bearing surfaces will come into contact with. Aerodynamic bearings are often referred to as foil bearings or self-acting bearings. Examples of this type of bearing include the read-write head flying over a spinning disk, crankshaft journals, camshaft journals, and thrust bearings for electrical generator turbines.

Aerostatic bearings

Aerostatic bearings require an external pressurized air source. This air pressure is introducedbetween the bearing surfaces by precision holes, grooves, steps or porous compensation techniques. Because aerostatic bearings have a pressurized air source, they can maintain an air gap in the absence of relative motion between the bearing surfaces.

Orifice and Porous Media Technology

Air bearings are typically classed as 'orifice' or 'porous media' bearings. In orifice bearings, the pressurized air is supplied to the bearing surface through a small number of precisely-sized holes. The concept is similar to the air hockey table game, but with the holes in the puck rather than the table. Porous media air bearings are quite different in that the air is supplied through the entire surface of the bearing (Figure 6). The porous material controls the airflow in the same way an orifice bearing would do if it had millions of miniature holes across its surface.

Broadly speaking, there are two techniques for achieving the compensating effect in air bearings. Orifice compensation is traditionally the most widely used method, but porous surface compensation is rapidly emerging as the method of choice due to its many advantages and increasing availability. In traditional orifice compensation, the precisely-sized orifices are strategically placed on the bearing and are often combined with grooves to distribute the pressurized air evenly across the bearing face. However, if the bearing face becomes scratched across a groove or near an orifice, the volume of air that escapes may be more than the orifice can supply, causing the bearing to crash even with normal air supply pressure. Rather than the small number of orifices of conventional air bearings, porous air bearings control the airflow across the entire bearing surface through millions of holes in the porous material. Because of this, they are harder to clog and will still fly even after being severely scratched. Even under normal operations, the air in an orifice air bearing loses pressure and creates pressure gradients in the air gap as it expands away from the orifice or groove. This is not an issue for porous air bearings where the air pressure remains almost uniform across the entire surface.

Figure 6 - Porous Media

Porous carbon has been found to be one of the best materials for this purpose, producing an ideal supply of uniform air pressure across the face of the bearing while automatically restricting and damping the air flow simultaneously. The carbon surface also provides greater bearing protection in the event of an air supply failure, and allows the bearings to be moved during air failure without damaging the support surface.

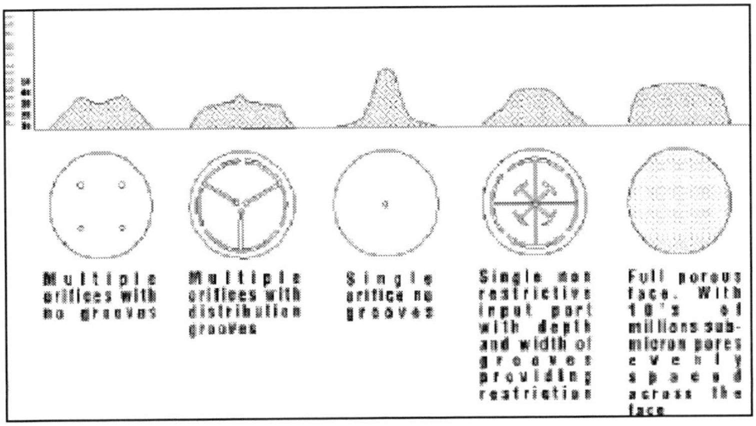

Figure 7 - Air Bearing Pressure Profiles

Lift-off and Collapse

When grounded, a flat orifice bearing only has the area of the orifice (and air distribution grooves) available to establish initial lift force. This limits the ability to preload orifice type air bearings. Porous media air bearings supply air pressure across the entire face, even when grounded.

Pitch moment stiffness

It's interesting to note how this effect distinguishes porous and orifice air bearing types when comparing pitch moment stiffness of an individual bearing. As orifice bearings are dependent on flow across the face of the bearing (e.g., an angular change in the gap), when one side of the gap gets large and the other side gets small, an unstable situation will be created where the available flow will rapidly lose its pressure following the path of least resistance (the large gap) and away from the area that needs it most (the small gap). The porous bearing, in contrast, still has pressure issuing from the small gap area. This gives the porous bearing higher tilt moment capacity and stiffness.

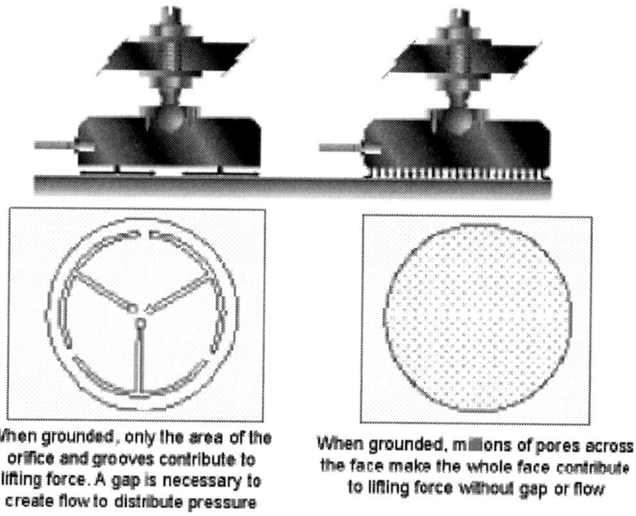

Figure 8 - Orifice vs. Porous Media

Figure 9 - Pitch Moment Stiffness

5. Air Bearing Products

Air bearings are typically available in five different types. Complete details on each type of air bearing product, including selection, mounting, configurations, and handling are covered in the later sections of this application guide.

Flat Bearings (Pucks)

Flat bearings may be round or rectangular, and are typically mounted using a threaded ball screw. They can also be bonded in place using a patented vacuum replication process.

Air Bushings

Air bushings offer the least expensive method of utilizing air bearing technology. They are designed to fit standard shaft sizes, and can often be used as direct replacement of existing bushings.

Vacuum Preloaded Bearings (VPLs)

Vacuum preloaded bearings (VPLs) are similar to flat air bearings, except they utilize vacuum to provide preloading of the bearing against the guide surface.

Air Bearing Slides

Air bearing slides are complete slide assemblies that use flat pad air bearing technology integrated into a compact air bearing slide package.

Radial Bearings

Radial bearings are ideal for very large rotary bearing applications and give you all the benefits of air bearings: no friction, no wear, no lubrication, and no noise.

6. Air Bearing Applications

Air bearings are used in a variety of applications including: coordinate measuring machines, precision machine tools, semiconductor wafer processing, medical machines, optical lens production equipment, digital printers, lithography, precision gauging, diamond turning machines, materials testing machines, crystal pulling, rotary tables, spindles, and friction testing. Air bearings offer distinct advantages for different industries and markets as outlined below.

Machines for the image setting industry

This includes machines for computer-to-plate (CTP) and high-resolution scanning in the prepress industry. These machines employ high speed spinning optics traveling down the center on an internal drum. A laser beam is reflected off the optic and directed to the internal diameter of the drum. Air bearings are important for axis of rotation and translate the spindle down the center of the guide. Pitch and yaw errors are magnified by the distance from the optic to the internal drum surface by the Abbe Principle. Air bearings are used in order to minimize these pitch and yaw errors which result in banding artifacts in the image.

Coordinate Measuring Machines

Most coordinate measuring machines (CMMs) are built with air bearings because they allow for infinite resolution. Because air bearings actually float on a pressurized film of air, there is no physical contact. This means only the shear of the molecules contributes to friction. The static and dynamic coefficients of friction at startup are identical, and there is no stick-slip effect. This minimizes lost motion and reversal errors around the triggering of the probe. Also, because air bearings are more repeatable and smoother than rolling element bearings, error correction is more effective. Mechanically, this allows for infinite motion resolution (putting the controls engineer back on the hot seat).

Testing Equipment

Many tensile and friction testing machines can be influenced by the friction in rolling element bearings. Wear in the bearings can also result in inconsistencies with testing processes. For this reason, many of the most accurate friction testing machines use air bearings to eliminate mechanical contact friction. Many testing machines require very accurate force control. The elimination of friction dramatically increases resolution of the instrument. Fatigue testing (another oscillating type of test equipment) will often cause 'fretting' in rolling element bearings. Being non-contact, air bearings are insensitive to high frequency, short travel applications.

High Speed Equipment

Machines are being designed today that have moving elements that may cycle as many as one billion times per year. It is simply not reasonable to attempt accelerated life testing on suchmachines. Another alternative is to switch the mode of wear by changing the bearing technology from roller bearings to air bearings. In air bearings, the speed or distance the bearing travels does not affect wear. The mode of wear in air bearings is erosion, so the amount of particulates in the incoming air is the determining factor of how long a bearing will last. Even assuming that relatively dirty air is used, the calculated life of an air bearing is measured in centuries, regardless of whether it is moving at one billion cycles a year or remaining stationary. The dynamic coefficient of friction increases with speed and will only contribute to heat problems at speeds of over 20 meters per second and then only in confined rotating applications.

Ultra accurate machine tools

Many of the most accurate machine tools in the world employ air bearing technology. The zero static coefficient of friction allows for unmatched performance during stage reversal in contouring applications. Very accurate velocity control and elimination of perturbations in the stage movement allow for the turning of optical quality surface finishes that are measured on the angstrom level. Errors in geometries on manufactured parts are often on the order of several millionths of an inch.

Linear Stages

The benefits of air bearings, and porous media air bearing technology in particular, are oftenincorporated into high performance linear stages. New Way® modular air bearings allow ultraprecision, frictionless stages to be designed and built with standard off-the-shelf components that are inexpensive and easy to use.

OEM

Air bearings are found throughout OEM precision machine applications. OEMs can either incorporate complete linear stages into their machines, or they can integrate their own stages using modular air bearing products.

Custom Projects and Test Rigs

If you thought that air bearing technology was beyond the scope of your one-time custom project, then think again. New Way is the only company to offer standard, modular air bearing products that can be easily integrated into your custom machine, test rig, or custom gauging application. With a wide selection of flat air bearings, air bushings, vacuum preloaded bearings, and air bearing slides in stock, the benefits of porous media air bearing technology are available to those with even the smallest of projects. The frictionless properties of air bearings provide for finer resolution of motion and allow significantly lower forces than can be achieved with rolling element bearings.

CHAPTER 14

AIR BEARING TURBOCHARGER

Creating a Turbomachinery Revolution

Oil-Free Turbine Engine

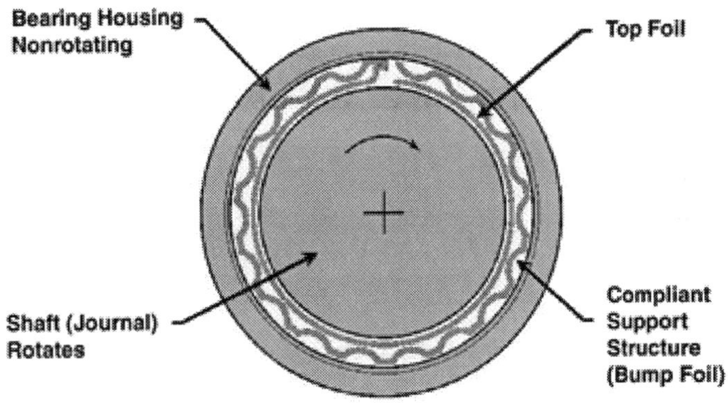

Image right: View of foil air bearing showing the non-rotating bearing housing, top foil, compliant support structure (bump foil), and rotating journal shaft. Credit: MACX

Everyone assumes that engines require oil for lubrication. We are defying traditional concepts of lubrication. Work is progressing on a revolution in the use of oil-free bearings that will eliminate the need for an oil-lubrication system in low-temperature, high-speed rotating machinery. The foremost challenge for this technology is the design of an oil-free turbine engine to power 21st century aircraft.

Bearings are used between the rotating and stationary parts of turbomachinery. Ball and roller bearings are commonly used in turbine engines. These bearings require an oil-lubrication system that provides a thin film of oil between the moving parts of the bearing. Without oil, the metal-to-metal contact would cause the engine to grind to a halt. Use of oil-free bearings removes the need for the oil system, thus, reducing weight, maintenance, and complexity of the engine. Studies have shown a potential reduction of 15% in engine weight.

Oil-free operation is attained through the use of foil air bearings. These bearings support a rotating shaft with two main elements: a top foil and a bump foil. The top foil provides a scuff-resistant start/stop surface and captures air to support the bearing. The bump foil provides support for the top foil and can be tailored for the required stiffness of the bearing.

This technology has been used for a number of years in low-temperature applications from magnetic disk drive read/write heads in PC's to the air cycle machines that maintain the pressure in commercial aircraft cabins. Recent advances in bearing design, low-temperature surface coatings and computer modeling are enabling new applications of these bearings.

Foil air bearings are self-acting hydrodynamic bearings that use ambient air as the lubricant instead of oil. Both air and oil are fluids that can perform the job of separating moving metal parts. The very high rotational speeds of turbine engine shafts (>10000 rpm) allow the replacement of oil with air. At these speeds, foil air bearings maintain the air film between moving parts by actually pumping air between the rotating and stationary surfaces. Air is drawn in and sticks to the bearing surface, no matter how fast it is moving.

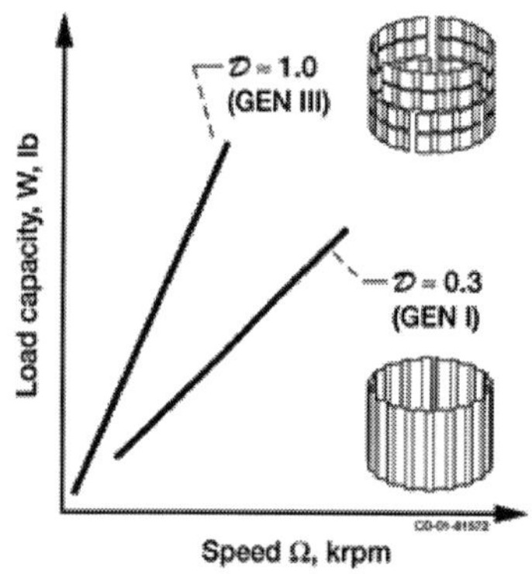

$$W = \mathcal{D}(L \times D)(D \times \Omega)$$

Where

- W is the maximum steady-state load that can be supported, N, lb
- $\mathcal{D}$ is the bearing load capacity coefficient, N/(mm^3•krpm) (lb/(in.^{3}•krpm))
- L is the bearing axial length, mm, in.
- D is the shaft diameter, mm, in.
- Ω is the shaft speed in 1000 rpm, krpm

Image left: Diagram showing load-carrying capacity of early and current designs of the bump foil. Credit: Magnetic Air Cars™

The pumping of air between the moving surfaces creates an air pressure that generates a load-carrying capacity. The trapped air and its cushioning effect act just like the air in a car shock absorber. These bearings provide a stiff, shock-tolerant support for rotating machinery. At high speeds, a very thin layer of air, just less than 1/1000" thick, can support hundreds of pounds.

The weight that a bearing can support is represented by its load capacity. For a foil air bearing, the maximum load capacity is dependent on the design of the top and bump foils, diameter, and length and speed of the rotating shaft. A faster shaft speed increases the load capacity by pumping more air between the top foil and the shaft. Larger bearing size also increases load capacity by providing a more bearing-supporting area. The foil design has a different effect. It determines how quickly the load increases with speed for a particular bearing. Early designs used a uniform corrugated bump foil (GEN I). Air leakage at the edges of the bearing caused it to have a relatively low maximum load capacity. Current designs (GEN III) are more intricate with multiple slits in the foil and changes in the number or spacing of bumps across the foil. These design features are better at trapping and maintaining the air film between the shaft and top foil. These refinements have resulted in great improvements in the weight that can be supported by foil air bearings.

Current designs are capable of supporting a pound of load for each inch of bearing diameter and square inch of bearing projected area for every 1000-shaft rpm.

Designers must tailor features of the bearing for a specific application. The stiffness of the backing bump foil is varied to maintain the thin compressed film of air and reduce air leakage out the sides of the foil. This will set the load-carrying capacity of the bearing. Also, high rotational speeds can make the shaft vibrate much like having an out-of- balance tire. These vibrating modes can be predicted with a rotor dynamic model. This computer model is used to predict the forces on the bearing and the motion of the overall shaft-bearing system. With changes in the shaft-bearing diameter, length, and thickness and the location of the bearing on the shaft, these vibration modes can be eliminated. Computer models let designers try out the design without going to the expense of building and testing actual hardware.

Traditional ball and roller bearings have rotational speed limits. At high speeds, the balls and rollers are forced away from the shaft and place large stresses on the bearing. Foil bearings do not have this limitation as there are no rotating parts of the bearing except the shaft.

Foil bearings now excel where low temperature and high rotational speed bearings are needed. The turbocharger test showed the bearings are candidates for operation in the hot section of turbine engines without the need for cooling.

Turbocharger companies are currently considering using foil bearings, with the low-temperature coatings, for possible use in **automobiles, trucks, boats and aircraft engines**.

This work on oil-free turbomachinery typifies MACX's efforts and will enable a turbomachinery revolution.

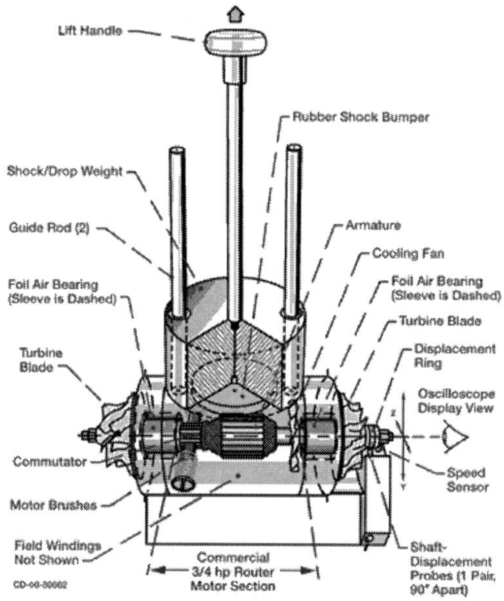

Image above: Oil-free foil air bearing shock tolerance

The application of oil-free bearings, such as air or magnetic types to automotive turbochargers, is expected to realize a reduction in mechanical losses while eliminating oil consumption. As a result, engines fitted with oil-free turbochargers will offer an improved response and lower fuel consumption, as well as reduced exhaust emissions. Toyota has designed compliant foil air bearings with uniquely shaped dampers for the journal and thrust bearings of small-sized turbochargers.

First, Toyota undertook a rotational test of a prototype turbocharger with air bearings. The results of this test revealed the need to increase the load capacity of the thrust air bearing relative to that of the journal bearing.

To solve this problem, they used 3D computational fluid dynamics to analyze the effects of the film thickness distribution between the bearing and runner surfaces of the thrust air bearing on the generated pressure distribution, in an attempt to increase the load capacities of the bearing.

Numerical analyses revealed three effective methods of increasing the load capacity, namely, increasing the size of the fluid charge in the bearing, generating the maximum pressure of the center of the bearing surface, and preventing the leakage of the fluid in the radial direction of the bearing surface. To realize these three improvements, we devised a new thrust bearing design with a shallow squared groove leading from the leading edge to the center of the bearing surfaces of the topmost foil.

The calculated pressure distributions on the surfaces of the thrust bearings both with and without the groove allows fluid to enter the bearing surface over a wider area and increases the amount of fluid by 70% relative to that without the groove. Because the groove does not link the circumferential edges or interrupt the center of the bearing surface, there is basically no radial leakage of fluid in the groove and the maximum pressure is generated at the center of the bearing. Numerical analyses with the grooved foil bearing indicated a 1.5-times increase in the maximum pressure and a 2.5-times increase in the load capacity relative to the conventional bearing. Figure 4 shows a trial version of the improved bearing together with a conventional version, both of which were installed in a turbocharger and then evaluated experimentally. A turbocharger with the improved bearing has been run at a rotational speed up to 200,000 rpm.

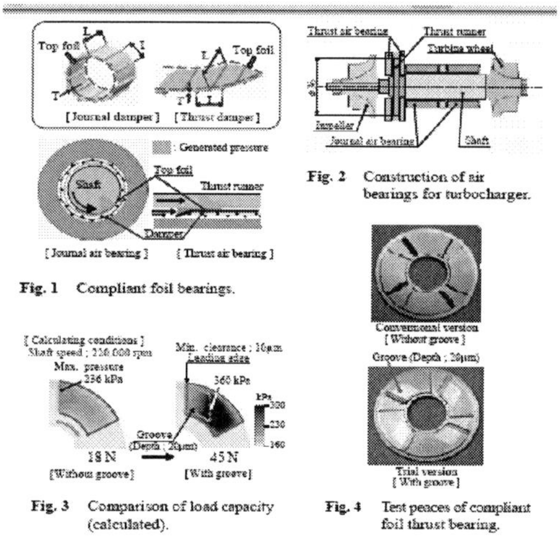

Mohawk Innovative Technology, Inc. (Miti)

It may run in a virtually frictionless environment with no oil required. It may run on compressed air. The turbocharger may compress air with five times less energy than required with a like model. This is a turbocharger that has been in development for five years. It runs on compressed air. The turbocharger may compress air with five times less energy than required with a like model using exhaust energy. It is a non-adiabatic machine, pure isothermal low delta T machine.

We believe we may be able to store excess power in batteries. We believe that our concept may be used with existing generators with a variety of turbines. We believe that our concept is more efficient than steam, internal combustion or coal burning methods. We believe as long as our generator and turbine have the capacity to convert the torque, volume, and the velocity that air-flow pressure creates, that the only restriction to power output is the pounds per square inch (PSI) rating of the air compressor. We believe that our magnetic engine may be designed for even larger power output.

We believe all future automobiles may be powered with this concept. One of the biggest problems facing the magnetic engine and air flow design industry today has to deal with heat and friction. We believe that we may have the solution to this problem. We believe that by utilizing lightweight materials along with air bearing design modifications, and with the unique design of the air flow device itself, that we may have eliminated the heat and friction problem associated with magnetic engines and air flow devices. We call this methodology "ISO Thermo Expansion Turbine Design". By eliminating heat and friction from our design, we believe that magnetic engines and air flow devices may now be feasible for generating electricity and operating a transmission at a much lower cost than traditional methods.

Chapter 15

Recuperator

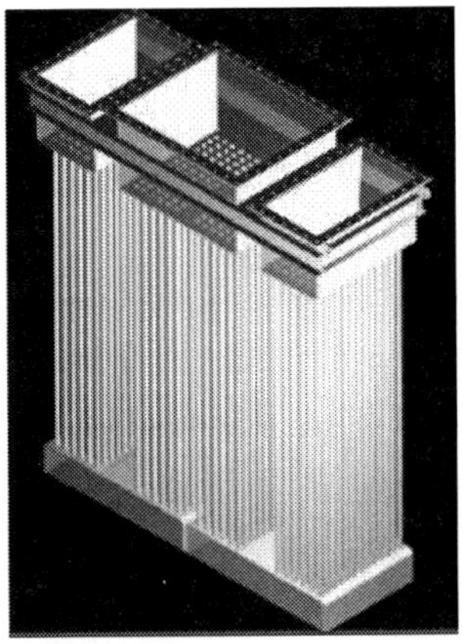

A recuperator is a special purpose counter-flow heat exchanger used to recover waste heat from exhaust gases. In many types of processes, combustion is used to generate heat, and the recuperator serves to recuperate, or reclaim, this heat in order to reuse/recycle it. The term recuperator also refers to compressed air counter-flow heat exchangers used for heat recovery in the highly pressurized air industries. By recovering some of the energy usually lost as waste heat, the recuperator can make an air bearing turbocharger engine or a compressed air turbine significantly more efficient. This recovery of the heat of compressed air significantly increases the efficiency of our air bearing turbine engine.

Chapter 16

Transmission

The patented *NuVinci* technology developed by Fallbrook Technologies Inc. (Fallbrook) is the most practical, economical and universally adaptable continuously variable planetary (CVP) transmission for human-powered and motor-powered vehicles and machines. The *NuVinci* CVP is ideally suited for applications in many major industries including bicycles, light electric vehicles, tractors, automobiles, trucks, and utility class wind turbines among others.

The *NuVinci* transmission uses a set of rotating and tilting balls positioned between the input and output components of a transmission that tilt to vary the speed of the transmission.

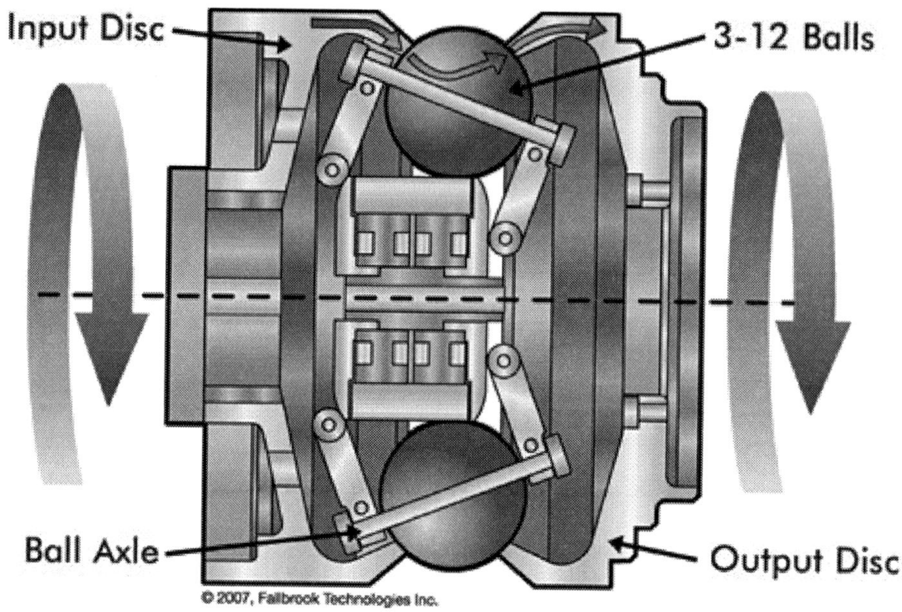

Tilting the balls changes their contact diameters and varies the speed ratio. As a result, the *NuVinci* CVP offers seamless and continuous transition to any ratio within its range, thus maximizing overall powertrain efficiency, with no jarring or shocks from the shifting process, and improving acceleration, performance and overall vehicle efficiency over conventional transmissions.

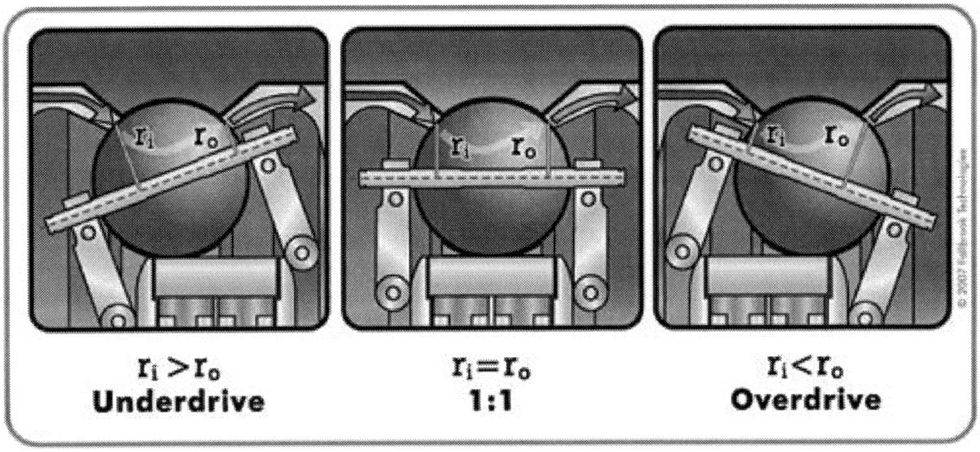

When compared to traditional continuously variable transmissions (CVTs), the *NuVinci* CVP is less complex, has considerably fewer parts, offers more stable control and scalability across product lines, is better packaged, and is less expensive to manufacture and assemble .

CHAPTER 17

DO THE MATH

Centrifugal Air Compressor

1. A centrifugal air compressor is a dynamic compressor which depends on transfer of energy from a rotating impeller to the air. Centrifugal compressors produce high-pressure discharge (psi) by converting ***angular momentum*** imparted by the rotating impeller (dynamic displacement).

In order to do this efficiently, centrifugal compressors rotate at higher speeds than the other types of compressors. Typically, half of the pressure developed in a centrifugal compressor is created by the blades slicing the air like a wing on an airplane, while the other half is created by slowing down the air flow leaving the impeller through a convergent-divergent nozzle (cd-nozzle). A cd-nozzle has a small area that leads to a big area. When the high speed air flow enters into the big area, the air slows down, thus giving up its velocity energy for pressure energy.

2. Angular momentum of a particle of a mass **m** (kilograms) with respect to a chosen origin **r** (meters) traveling at a velocity **m/s** (meters per second) is given by L = mass x velocity x radius, or **L= mvr**. The most important concept here is momentum being mass x velocity, but because we have a radius involved, we have to calculate its effect on that momentum. The objective of an impeller is to spin at such a high velocity that it can suck in air and throw it at a high velocity through its designed radius to the cd-nozzle. Again, half of the pressure is developed before the air leaves the impeller through pressure variation of the spinning blades slicing the air. A picture of what an impeller does is shown below.

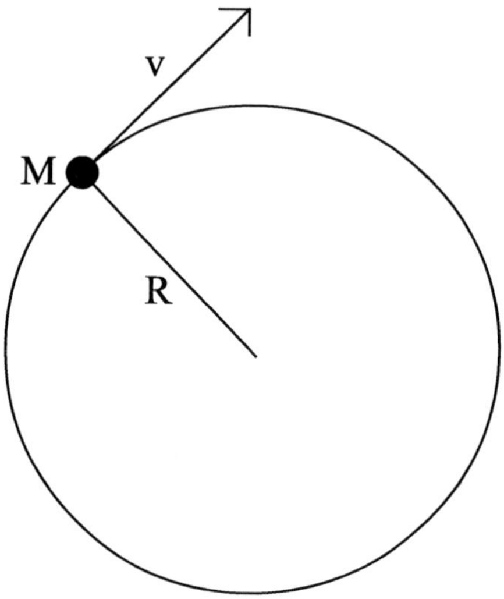

3. When air is compressed, it creates heat. According to the physics textbooks, the amount of heat that will be developed when air is compressed is determined by the formula **(T2/T1) = (P2/P1)**$^{[(g-1)/g]}$. What this equation states is that with any air compressor system, the temperature difference between the inlet and outlet of the compressor determines the pressure increase of the machine. A calculator below shows an example of a typical process.

Physics Compressed Air Outlet Temperature Calculator

Input Temperature	80°F	Ambient air temperature
Input pressure	14.7 psi-a	Atmospheric air pressure
Output pressure	26.46 psi-g	Compressor output pressure
Compressor efficiency	70%	Efficiency of the compressor
Output temperature	341°F	Pressure ratio 2.8:1

What is stated is that if air is compressed by a pressure ratio of 2.8:1 (26.46 psi-g) starting at 80°F to the inlet of the compressor, the air will come out at the outlet at 341°F if the compressor is 70% efficient. To determine how much energy was used to accomplish this, we need the mass of the airflow in kilograms per second. For example, if the mass flow rate is 0.19 kg/s, we can use a well- known physics calculator to determine how much energy was used to compress the air because the calculator above only calculates temperature rise. We need to know energy in joules using another calculator called the specific heat calculator.

The specific heat is the amount of heat per unit mass required to raise the temperature by one degree Celsius, which is given by Q = CM Δ T. This equation states Q being heat added in joules is equal to the specific heat of matter (C) multiplied by the mass in kilograms (M) multiplied by the change in temperature (ΔT). The specific heat of air in physics is 0.286 calories/gram Celsius, which is 4 times less than water. In other words, water takes 4 times more energy to heat up to a certain temperature than the same mass with air.

By using the specific heat calculator and physics compressed air outlet temperature calculator, we can determine the energy used to compress the air in the above example. To solve for Q, it is stated that for a specific heat of air being 0.286 (C) multiplied by a mass of 0.19 kg/s (M) multiplied by the change in temperature 341°F-80°F (ΔT), the total energy is 32,989 J, which is about 33,000 Joules, or 44.29 hp.

It must be noted that Celsius must be used and not Fahrenheit to get the right information. What this says is that to compress 0.19 kg/s of air at a pressure ratio of 2.8:1, 33,000 J worth of energy is required. An air compressor that requires this is manufactured by Garrett, and is called the T3-50 trim, which flows 0.19 kg/s at a 2.8:1 pressure ratio. The hot exhaust from the engine drives this air compressor, which is called a turbocharger.

The disturbing part of this is the temperature of the compressed air coming out of this turbocharger when running at 2.8:1 is 341°F, even if the ambient atmosphere is at 80°F. What engineers do is flow this hot compressed air through a heat exchanger which dumps all the energy (33,000 J) into the atmosphere. Shortly after, the air is still at 2.8:1, but at an ambient temperature of 80°F. Therefore, all the energy in the beginning used to compress the air is wasted through the heat exchanger! According to the Department of Energy, about 80% of the energy used to compress air is thrown away as heat, leaving the remaining 20% for actual end use. It is therefore MACX's objective to compress air while only requiring the 20% instead of the extra 80% which is lost as heat to the atmosphere. From the above example, instead of requiring 33,000 J to compress the air, MACX has developed a process to which only 20% of 33,000 J will be needed, which is about 6600 J.

However, according to current physics, this is not possible when looking at how engineers dump all the energy used to compress air to the atmosphere and still end up with the increased pressure, making it possible to compress air while generating the least amount of heat. MACX also has a working prototype that validates the mathematics. By using basic angular momentum concepts, it is proven that air can be compressed with only the 20% requirement as stated by the Department of Energy.

MACX'S Formula for compressing air while generating the least amount of heat, friction, and start up load

Mass x Velocity x Radius /1 sec = A (Newton)
A/Gravity = B (New Virtual Mass Flow Rate)
B x Velocity = C (New Momentum of the air)
C/Radius/Compressor Efficiency = Joules

Example

0.19 kg/s x 485 m/s x 0.03 m/1 second = 2.7645 N. A = 2.7645 N
2.7645 N / 9.8 = 0.282 kg. B = 0.282 kg (New Virtual Mass Flow Rate)
0.282 kg x 485 m/s = 136.81 N. C = 136.81 N
136.81 N / 0.03 m / 0.70 = 6514.97 J / 8.74 hp
approximately 20% of 33,000

It must be noted that the mass of an object is equal to a certain amount of force because gravity exerts a constant force on an object depending on how much matter it contains. It must also be noted that the original mass flow is 0.19 kg/s, but because we have circular motion with the radius of the impeller involved, it takes more work to move 0.19 kg/s at a velocity of 485 m/s than with linear motion.

Therefore, when calculating the radius, we get an angular momentum of 2.7645 N worth of force. Converting Newton back to kg requires the gravity of the planet, which is 9.8 m/s/s. As the above example, 2.7645 N/9.8 = 0.282 kg/s. Thus, 0.282 kg/s is now the ***new virtual mass flow rate***. MACX calls it virtual because the impellers' geometry when spinning at max speed can only flow 0.19 kg/s, but because we have the radius of the impeller involved, the flow is now virtually 0.282 kg/s and the force is now 0.282 kg/s x 485 m/s x 0.03/1 sec = 4.103 N.

What this means is that we must input 48.4% more energy into the impeller to move 0.19 kg/s at a velocity of 485 m/s. 0.282/0.19 = 1.484 because we have circular motion.

The T3 turbocharger example above requires 33,000 J. According to the textbooks, this is necessary, no matter what the machine. However, engineers do not take into account entropy generators in their air compressor systems. They typically make a turbomachine that has **heavy parts** to take the heat while being stable. They use **high-friction bearings** and they use **hot exhaust** to drive the turbomachine.

When testing the energy requirements, they use a jet engine to drive it up to max speed, measure the mass flow and the temperature difference between the inlet and outlet, as well as the pressure increase. From here, they make a **law** that this is what it takes to compress air. However it is not a law, but only the law of that particular machine. Entropy is waste heat.

As stated by the Department of Energy, 80% of air compression is waste heat but required, according to the laws of physics. This 80% waste heat is caused by entropy generators and not by compressing air. Entropy generators are friction, inertia, and high heat drive. If these entropy generators are reduced to a minimum, the 80% waste heat energy will not be needed to compress air to a high pressure, and instead, only the 20% will be needed. MACX has developed a process that puts entropy generators to a minimum, which allows compressing air requiring only 20% and not the extra 80% on top.

The problem with air compression

It is necessary to compress air while generating the least amount of heat. The heat generated in a traditional air compressor is produced by friction, inertia and high heat drive. The main areas of friction are bearing friction, oil friction, gear friction and high energy drive system friction.

If these aforementioned problems are eliminated or reduced, there will be more compressed air produced with reduced waste heat losses. As these problems are completely eliminated, we now have 5 times more available, efficient compressed air available with no extra energy needed.

Compressing air at the 20% end use requirement

MACX has researched, discovered, filed patent on how to use the compressed air coming from a traditional air compressor to generate more compressed air on the order of a 3-5x magnification. MACX accomplishes this by eliminating entropy generators, namely bearing, oil and gear friction, by using gearless air compressors and air bearings. Air bearings have 100-1000x less friction than traditional bearings. A gearless air compressor is a turbomachine which uses expanding fluid to drive the machine.

Another entropy generator is inertia. Compressors today have their operating components made of heavy metals, which require a lot of energy to turn and sustain. Once they are running, bearing, oil and gear friction fights to slow down the compressor to which constant high energy is needed to sustain the machine.

Since MACX is using air bearings and a gearless turbomachine, no friction is generated, therefore freeing up needed drive energy. The 3rd area of entropy is drive friction or heat change friction. Drive friction occurs in an electrical motor. As the motor spins to power the compressor, the motor's bearings and oil fights to slow down the motor. Also, a counter current is produced to slow the electric motor down as it spins, making constant high energy a necessity to sustain the entire electrical compressor. Heat change friction occurs in a gas turbine compressor. This type of friction occurs in the combustion turbine. The high temperature gases turn the turbine to drive the compressor and the heat of air compression becomes excessive. In other words, a turbine can drop 1000°F from inlet to outlet and the compressor will copy this energy by compressing air to the same level of energy, getting the air really hot. The objective of compressing air is to generate the least amount of heat possible during the compression process. However, if the gas turbine compressor has a hot turbine (as all do), it will also have a hot compressor. The area of drive and heat change friction are eliminated by our fluid flow amplifier device (FFA). The FFA is designed to work with electrical or fuel-powered air compressors. The FFA device can be powered by a traditional air compressor, enhancing its ability to produce more cost-efficient usable power.

How MACX compresses air

Cooled, compressed air coming from a standard electrical or fuel-powered compressor is sent to our fluid flow amplifier device (FFA), which is the primary air stream. As the compressed air flows through the FFA device, the temperature drops below ambient temperature at the nozzles and the air travels at approximately Mach 1 speed (760 knots), breaking the sound barrier. During this process, the primary air draws in ambient air on the order of 15-25x the amount of the primary air used and speeds it up.

The primary supersonic airflow gives its velocity energy to the ambient air and creates a mixed uniform primary and ambient airflow stream of about Mach 1 momentum. This new airflow stream is at a cool temperature and contains nearly all of the energy used to create the compressed air in the beginning but it is not hot. Considering a FFA device utilizes proven pressure-velocity exchange, according to physics, the new airflow stream eliminates heat change friction that occurs in gas turbine compressors because the turbine does not cool down.

The new airflow stream enters into our turbine at a temperature below ambient, and therefore, it cannot reject heat to the ambient, but only gain heat. The air goes into the turbine cold and comes at the same temperature, thus no heat change friction is generated. The new airflow stream can now drive a superlight weight turbine and compressor (carbon fiber) on air bearings with nearly no friction and heavy start-up load.

The compressor fan can now be made larger than the traditional primary compressor fan and produce 3-5x more compressed air for generating power, thus making traditional air compressors 3-5x more efficient. The advantage of the FFA device is that it takes compressed air from a traditional air compressor and it gives it to the ambient air, thus we gain massive amounts of ambient air traveling at approximately Mach 1 speed, which has the kinetic energy and ability to drive a low entropy turbomachine.

It should be understood that MACX's formula can be used to prove that the energy requirement of the T3 turbocharger is 33,000 J. The formula is shown again:

Using MACX'S formula, it can be proven that the T3 compresses air with no excessive energy required.

Mass x Velocity x Radius /1 sec = A (Newton)
A/Gravity = B (New Virtual Mass Flow Rate)
B x Velocity = C (New Momentum of the air)
C/Radius/Compressor Efficiency = Joules

The Fact. One must remember that the original mass flow of the T3 compressor is 0.19 kg/s. Using angular momentum concepts showed us that the ***new virtual mass flow rate*** at minimum is 0.282 kg/s because we have circular motion. Taking (0.282 kg/s x 485 ms/0.03m)/0.70 = 6512 J.

The T3 turbocharger produced by Garrett has a ***new virtual mass flow rate*** of 1.427 kg/s which is approximately 5 times the amount of the minimum value of 0.282 kg/s. Below uses MACX's formula to show how the T3 turbo requires 33,000 J (5x the amount of energy, which is equal to the 80% of the energy that is called 'waste heat').

0.9618 kg/s x 485 m/s x 0.03 m/1 second = 13.99 N. A = 13.99 N
13.99N / 9.8 = 1.427 kg. B = 1.427 kg (**New Virtual Mass Flow Rate**)
1.427 kg x 485 m/s = 692.56 N. C = 692.56 N
692.56 N/0.03 m/0.70 = 32979.50 J, which is rounded to 33,000 J.

As seen, the minimum virtual mass flow rate should be 0.282 kg. However, the T3 turbocharger raises it to about 1.427 kg. This extra energy is not necessary and is excessive waste. By eliminating or reducing entropy generators in the turbocharger, it becomes possible to reach the ideal virtual mass flow rate, which is 0.282 kg. Therefore, the MACX method is using less energy to produce the same results, making cost-effective air compression work, and also utilizing minimum effort to produce maximum results.

Math formula contributed by Elijah and Elisha Dumas

MACX Power and Electric Animation

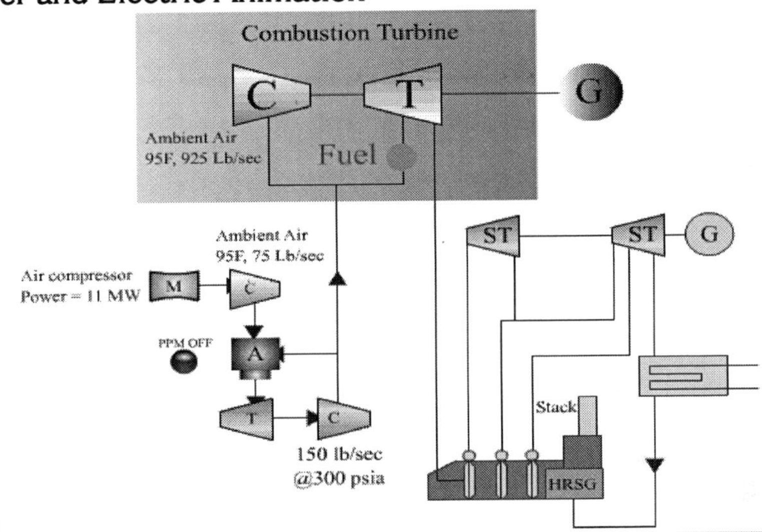

Quantum Physics

The output of one operation becomes the input of another (z = Z2 + C).

A person in a swing may weigh two hundred pounds and a weak boy pushing it may weigh but fifty and may only have the strength to push a pound. Yet, if he times his pushes to coincide with the turn of the swing from him, and keeps adding a pound each time, he will eventually have to stop to avoid hurling the occupant of the swing out into space.

"The principle cannot fail," Tesla would say. "It is necessary only to keep adding a little force at the right time." The problem was to give the particles a series of impulses that were a little stronger each time, until, like a child being pushed on a swing, the momentum was greatly increased.

Chapter 18

Gatekeepers

Gatekeepers are men or women in the scientific community. Their scientific validation may certify whether a new technology may or may not receive funding. These academic scientists make scientific validations of new breakthrough technology through feasibility reports from their direct observation and hard mathematical data.

The following is an example of a gatekeeper observation.

The Patent and the Free Energy Claim

We saw that the inventors closely associate this patent with investment in their free energy machine. Is this reflected in the patent itself? * The provisional of November 1998 makes no free energy claim, only listing the purpose as "to improve the efficiency". * Mr. Cliff Carew, the attorney who drafted the patent in November 1999, told me that the inventors presented him with the central idea of using a timing switch to control electrical energy. In his opinion, it is a valid method of improving the efficiency of an electric motor. They did, however, claim that it produces free energy, and he stressed to them that perpetual motion is impossible. * The patent text is very quiet on free energy, but pages 15 and 22 list among possible variations: *The output produced by the rotary device can be mechanical and electrical at the same time... Current generated in the stator coil windings can be used as an output, and likewise, the torque generated by the rotor can be used to supply a mechanical output.* * When I put this to Mr. Carew, he responded that this is meant in the same way that a car alternator may, at the one time, put out electrical energy (via the terminals) and mechanical energy (e.g., to its cooling fan). * My difficulty with this interpretation is that a car alternator has a third interface: a shaft driven by a belt, which undoubtedly puts in energy. In contrast, the machine described in the patent has only two connections to the outside world, electrical to these stator coil windings, and mechanical to the one shaft, so there is no possibility of energy input at a third interface. This is confirmed in a letter from John Christie dated January 4, 2001: *By reconfiguring the coil, we can pump power back to the battery source, which then holds or even increases in charge level as the motor runs.* Therefore, there can be both electrical and mechanical output (as distinct from input) at the same time. This indicates that the inventors intended the patent to support claims of free energy and perpetual motion. More recently, the inventors' web sites and media statements show that they are relying heavily on the patent application to support their business claims for generating free energy.

The Aspect of Critical Timing of the Electric Current

Putting aside the claim of free energy, does the patent contain any substance of value? The patent details describe at great length a special switch which provides electric pulses of specific timing and duration to the electromagnet. It shows a commutator with conductors that are tapered in order to fine tune the pulse duration by adjusting the brushes vertically. The writer apparently believed that by applying current just long enough to overcome the natural magnetic attraction between the rotor's permanent magnets and the coil's iron cores, and then removing the current, more energy can be produced than is expected. John Christie has described this to the media, by web site, and to me. Based on my experience as an electrical engineer, I see many flaws in the claim that this switch is novel or effective in improving efficiency: * Variable timing can be obtained electronically with ease, so this adds nothing new.

* More amusingly, the writers do not seem to realize that in breaking the battery-electromagnet circuit instantly, the inductance will cause the current to keep flowing. In fact, copious arcing is very visible in their video of the device running! This wastes energy and will quickly erode the contacts. * They believe this timing is especially critical, as the switches are to be closed for precisely 12 degrees, 51 minutes and 50 seconds of angle (They could not possibly measure to this accuracy, so mathematicians among our readers might like to reverse engineer this figure to see what theory predicts it)! Interestingly, the provisional patent a year earlier listed: *typically, 16 degrees 30 minutes of arc, the pulse being applied at 20 to 40 minutes of arc beyond a predetermined position...* * They believe this special timing makes the generator more efficient.

Free energy?

More output than input. The patent description includes a diagram of waveforms, but I cannot understand it. On seeing my comments, Mr. Cliff Carew responded:

> *The idea of combining an alternator with an electric motor by using a timing switch to control electrical energy fed to the overall system seems to be a viable proposition and is the one which is covered in the patent specification. Interestingly, the inventors are vigorously promoting it only as a free-energy machine, and its* **real** *efficiency is about 33%.*

Other Errors

The patent contains errors of spelling, grammar, and figures not referenced. For those readers with some electrical knowledge, the following should be of interest: * Page 3 states, "... the back EMF urges the rotor to rotate ..." Actually, a back EMF provides no force unless there is a circuit to allow current to flow. Page 19 repeats the mistake "... repulsion produced by the back EMF ..." * Figure 7, showing input voltage vs input current, and Fig 8, showing natural magnetic attraction, etc., do not make sense from an engineering point of view. * The writer of the patent believes that continuity of forward force is essential. He states that the coil should be energized at the moment the resultant of other forces becomes opposing (p. 3), and should be de-energized at the moment that other forces become assisting (p. 2). Without this, the rotor will not be able to spin continuously (p. 11). This belief ignores the inertia (flywheel effect) of the rotor. * **A real engineering description of a motor would include analysis of ampere-turns, magnetic reluctance, air gap flux and the like**. This patent instead contains confused ideas like "natural magnetic attraction", "induced induction", and "ratios per second". * The magnetic force is referred to as the "mug force" (A Freudian slip perhaps, referring to investors?) * There are many irrelevant digressions.

Page 13 expounds that the coils may be wound from copper, silver, aluminum or other metals. The shape of the winding wire may be round, square, triangular, rectangular and others, and so on. In fact, the magnetizing force of a coil is simply the number of turns of wire, multiplied by the current flowing; the choice of metal and the shape are irrelevant to the principle of operation. Mr. Cliff Carew responded that my criticisms are just "wrong interpretations placed on wording that has been used". My view is that the patent documentation shows a lack of sound knowledge of mechanical and electrical principles.

Can Patent Perpetual Motion Machines be Patented?

It is often reported that in the USA, you cannot patent perpetual motion machines. This is due to a statute for prohibition if the patent violates the laws of nature. The Australian commissioner for patents, Ms. Vivienne Thoms, advised me that Australia has no such statute, so that the government examiner cannot use that basis to reject a patent. However, the applicant must describe how it works, and give the best method for performing the invention. If the patent is later challenged in court, and the plaintiff shows that it violates the laws of nature, then the court will accept that the invention cannot work, and will not be useful. This is grounds for revocation of the patent.

Will Attorneys cater to Perpetual Motion Machine inventors?

I posed this question to Mr. Cliff Carew of Griffith Hack, who drafted this patent. He wrote that he understood it as a valid method of controlling a machine, which provides improved efficiency. *I believe in the laws of physics, and that there is no such thing as a perpetual motion machine and I have made this clear to both John and Lou.*

In any case, based on other patents I have seen around the world, it is clear that there are some patent attorneys who are willing to draft and file patent applications which violate the laws of nature. Their staff (if competent) must know that the subject device will never work, yet they proceed to charge fees to the applicant. They could claim that they are supplying a service which the applicant desires, and so there is no victim. Skeptics would argue that this often indirectly results in investors being cheated out of their savings. The ethics of the situation need to be investigated.

Comparison with Other Perpetual Motion Machines

Patents for perpetual motion machines cover a wide span of complexity. My collection includes the following:

* The most fundamental such machine I have seen was a German *(Patent Fig 11, Waveforms)* patent showing a car motor containing eccentric weights on a shaft, the arm length varying with rotation angle. This scheme is re-invented periodically, such as in an Australian bicycle where the pedal arms are lengthened during the down-stroke. A rudimentary education in mechanics would reveal that this principle cannot produce free energy *or* improved efficiency. * Of similar simplicity, but employing magnets, a device I saw at an exhibition consisted of two concentric aluminum cooking pots, the inner one (the rotor) free to spin on bearings. The inventors had glued magnets on the facing surfaces of the pots. When they gave the rotor a spin by hand, it lurched encouragingly. Even at the exhibition, they were desperately rearranging the magnets, in the naive belief that once they found the right pattern, it would continue accelerating. * The Lutec machine is more complex being electromagnetic, but nevertheless, the theory and practice is well-known. **The Lutec patent contains no x-ray generators, cold fusion cells, or other high-tech devices** which could gain access to exotic energy sources.

* The most complex perpetual motion machine I have personally investigated is arguably the monster machine which included shaft power, electricity generation, heated gases, oscillating pistons, and mechanical rectification by ratchet. It cost a fortune to make, but to the immense puzzlement of the inventors (who were skilled Italian technicians), it refused to run.

Why can't Free Energy Machines work?

This section reflects my many attempts to explain some basic physics to proponents of perpetual motion machines. I address it to the many I have argued with, to those proponents who may have a chance to read this, and to skeptics in dealing with such inventors they may come across.

Energy exchanges on Different Scales

There is a quantity called energy, which matter can possess in many forms, including: * Gravitational (more when a mass is high);

- Kinetic (when a mass is moving fast);
- Heat (more when the material is hot);
- Chemical (more when a material has the power to burn or go bang);
- Electrical (carried by current, and stored in electric fields);
- Magnetic (stored in magnetic fields);
- Nuclear (when an atom's nucleus undergoes fission or fusion).

The basic laws of physics are very well established. That is why we can successfully design jumbo jets and cell phones. The most fundamental laws describe the forces and motions (interactions) of elementary particles. The laws show that in every interaction, *without exception,* the energy given up by one particle is transferred to another particle. This is equivalent to saying that in this basic interaction, the total energy is constant. This has been verified in a billion experiments, in classrooms and laboratories around the world. All scientists and a majority of members of the human race accept it. When the particles are grouped together to form atoms and molecules, many of the fundamental interactions can be grouped together. From there, higher-level interactions and laws come into play (for example the field of chemistry). A chemical reaction can be regarded as summations of fundamental interactions. Thus, the total energy should be conserved, and indeed, measurements and practice confirm it. When bulk materials are fashioned into a machine, the laws can be further coalesced into so-called macroscopic laws, which take into account the averaged properties of the materials. An example might make this clear. Consider a rotating electromagnetic machine connected to a battery. Engineers design and test such systems every day. **Interactions include:**

- Mechanical forces between two parts (due to physical contact, e.g., bearings);
- Inertia (energy stored in the moving parts);
- Chemical to electrical (electrolyte and electrodes in battery generate current);
- Electrical to magnetic field (e.g., current flowing from battery to coil of electromagnet);
- Magnetic fields (e.g., surrounding permanent magnets);

Originally in the Skeptic *Vol 5, No 3*

Free energy?

- Magnetic to mechanical (e.g., magnetic field generates torque urging rotor to rotate);
- Magnetic to electric (e.g., moving magnet in generator produces EMF (voltage) in a coil);
- Electrical to heat (e.g., coil heats up).

When such a motor or generator system runs, what do we find out about the energy? With so many exchanges, surely there is opportunity for new phenomena? Many free energy proponents believe (or at least claim) that by carefully fashioning the magnets, coils, pole pieces, etc. (or the equivalents in their field), they can create more output than input. Sorry, both the theory and the experiment are again unanimous: **the total energy is conserved**.

Analogy with the Economy

Occasionally, I try to explain by analogy. Consider two people with coins in their pockets. In an interaction, say A gives B $5 (in exchange for a bag of apples). As a result, A has $5 less and B has $5 more, and the total money held by the pair is unchanged. This is the same idea as exchanging energy in an interaction between two elementary particles. Now, consider a payment from one company (group of people) to another. The payment can be regarded as the combination of many elementary transactions, and the company payment is the summation of the individual transactions. It follows that money is conserved also at the larger scale. Finally, consider transferring money among a set of companies; will this break the conservation law? Alan Bond tried this; his round-robins of loans and borrowings appeared designed to generate money out of thin air. He could not, or would not, see that the total money is constant. Most people can see that, no matter how complex the purchases and payments, they are merely a summation of more basic interactions and that there can be no "net gain". I hope that readers in doubt can see that the workings of machines is analogous to the economy. Energy is conserved at the lowest level, and through summation, it follows that it is also conserved at the highest level. The law of conservation of energy is also known as the first law of thermodynamics.

A Primer in Power and Energy

When confronted with claims of energy-saving devices, it is useful to be able to tell fact from fiction. Knowledge of some terms and units might help. **Power** is measured in *watts*. A light bulb consumes around 100 watts of electricity, and a car engine might produce 110 kilowatts of mechanical power at its output shaft. **Energy** is the power totaled over a certain time interval, and is measured in *joules*, one joule being one watt for one second. Our electricity bill charges us for electric energy in kilowatt-hours (kWh), and our gas bill charges us for the chemical energy which is available as heat output, in megajoules (MJ). A calorie in food is about 4.18 joules of chemical energy available to our metabolism. The first law of thermodynamics simply states that energy is conserved, i.e., may be converted between forms, but may not be created or destroyed.

There is also a second law of thermodynamics, which states that in any transfer of energy in a machine, some is wasted as heat. This means that if we put in say 100 watts of electric power into a motor, we can only get less than 100 watts out as useful shaft power. The ratio is called the **efficiency**. Machine designers have been striving for hundreds of years to make their machines more efficient. Consider car engines, batteries for mobile phones, power stations, and aluminum refineries. Any designer who can gain 1% in efficiency for their process (say, from 91% to 92%) is doing their job well. The above comments address common energy sources such as chemical, electrical, magnetic, and mechanical. There are some exotic energy sources which should be mentioned for completeness. If an inventor claims his engine runs on:

- Nuclear fission (e.g., from uranium);
- Nuclear fusion (e.g., Horvath car and cold fusion);
- Antimatter;
- Quantum fluctuations;
- Zero-point energy;
- Wormholes;
- The electromagnetic aether; then we cannot immediately dismiss their claims on the basis of conservation of energy alone. A deeper analysis (usually by a specialist) is called for, and in many cases, we find that the claim does not ultimately make sense because the inventor's technical knowledge was deficient or feigned, as was the case for the Horvath hydrogen fusion car.

Gatekeeper's Conclusion

Australian skeptics (and their counterparts around the world) will continue to investigate claims of free energy, from both the theoretical and measurement points of view. We will keep an open mind on each case until the evidence is in.

Meanwhile, we can categorically say to such free energy proponents that, according to the laws of physics, their machines can't work. All measurements to date have confirmed this. Thus, such claims are based on false premises, and fall under the description of scams. To seek money or advantage from them, risks being labeled fraudulent. To people considering investing in such schemes, we suggest you say to the promoters "Australian skeptics say it is **impossible**, have you accepted their challenge?"

The Gatekeepers claim that they will keep an open mind but in fact they believe that it is impossible. The Gatekeepers represent the envious industries, the government, universities, scientists and critical journalists. They are the major cause of all travail on this earth. They have no imagination and do not believe in anything. They only believe in what can physically be observed in the present and have no vision for our future. I have clearly witnessed that many so-called scientific intellectuals in high positions of influence are FRAUDS. Why? Because they have the mentality of a parrot. They have simply memorized the great contributions of important scientific innovators of the past and can quote those contributions like a parrot. Yet at the same time, they have absolutely no true understanding or **curiosity** regarding the subjects they quote.

The energy crisis in this world – and the tragic deaths of hundreds of millions of people worldwide because of their abject poverty due to a lack of cheap energy for a modern national economy – rests directly on the shoulders of our scientific community, and specifically on the shoulders of our electrical engineering community and profession. In fact, the technology in this book can be used to provide abundant, non-polluting, inexpensive electromagnetic energy to power the world.

- No more water or energy shortages - anywhere on the planet.
- No more fossil-fuel-related pollution.
- No more skyrocketing energy costs.
- No more global warming caused by the use of hydrocarbon fuels.
- No more nuclear reactor wastes.
- No more coal-burning power plants.

This technology will provide abundant, inexpensive, non-polluting energy to power homes, factories, automobiles, ships, planes, appliances, etc. In short, the entire world can utilize this revolutionary technology.

- The Three Laws of Thermodynamics were conceived without an understanding of the relationship between heat (gyroscopic massergies/electromagnetic energy) and matter.
- The Three Laws of Thermodynamics were conceived without an understanding that there is an energy relationship other than the simplicity of [Work = Force X Distance], [Power = Work/Time], and [Force = Mass X Acceleration].

- The Three Laws of Thermodynamics were originally conceived without any knowledge, understanding, or anticipation of Einstein's equation of $E=MC^2$.
- The Three Laws of Thermodynamics were originally conceived without an understanding of gravity, electricity, magnetism, inertia, matter, and planetary motion.

QUESTION: If none of these things were understood at the time that the Three Laws of Thermodynamics were conceived, how can these three laws be so "all encompassing" as to be capable of predicting --- on a seemingly "infallible" basis --- the "Doom of the Universe" and the "Total Impossibility of Perpetual Motion?" Those who made such predictions must have understood the mechanical workings of the entire universe. It is easy to recognize that in this sense, these laws have operationally been a deliberate attempt to close young minds that would be otherwise willing to question the "finality" of the Laws of Thermodynamics.

"There is NO PLACE in science that negativism should be allowed to exist! The entire history of science has proven repeatedly that whenever it has been thought that something was not possible, it later turns out to be possible. Therefore, as the facts have proven, science should put forth positive statements of hopes and dreams that will perpetually stimulate the creative processes of the human mind. In contrast, throughout my sincere, scientific efforts of nearly two decades, I have had to fight against many negative "scientific statements" that were and are wrong. Such injustice has not been unique to my efforts, but, on the contrary, it has been the common fate of most creative individuals throughout the history of science......" *Joseph Newman*

All such successful self-powering asymmetric permanent magnet motors have been ruthlessly suppressed to date. The latest highly suspicious "capture" of a legitimate procedure to make asymmetric permanent magnets – and thus suppress the process from use by our citizens – was by our own U.S. Department of Energy in 2001. Apparently, the DoE has also kept this fact – the availability of the mass production of asymmetric permanent magnets with high energy products, and thus, the easy assembling of self-powering asymmetric PM motors and motor generators – is completely unknown to the President, the Congress, and the American people.

In short, turn all our young doctoral candidates and post-doctorates loose on the problem of developing and using practical asymmetrical Maxwellian systems, and in two years, there will never again be another energy crisis anywhere on this planet.

CHAPTER 19

CLEAN ELECTRICAL ENERGY FROM THE ACTIVE VACUUM

Tom Bearden
Jan. 2002

Introduction

No electrical engineering department in the Western World presently teaches what powers an electrical circuit, or what actually powers the electrical power grid. None ever has. It also does not appear in a single electrical engineering textbook in the Western world, nor has it ever appeared in one.

All the hydrocarbons ever burned, nuclear fuel rods ever consumed, steam turbines turned, and generators rotated, have not directly added a single watt directly to the external power line and to the power grid. Nor has any windmill, nuclear power plant, battery, or hydroelectric generator or solar cell array. None ever will.

Every electrical circuit ever built — and those built today — are in fact powered by electrical energy extracted by the circuit dipolarity from the local seething vacuum, from active space itself. But our engineers are trained to build circuits which also self-destroy the extraction of that vacuum energy, faster than they power their loads.

The leaders of our scientific community — including the National Science Foundation (NSF), National Academy of Sciences (NAS), and National Academy of Engineering (NAE) — are unaware of what actually powers an electrical circuit or the electrical power line. Scientists have not integrated into classical electrodynamics and electrical engineering — and in their own thinking — the *broken symmetry of opposite charges* — such as of a common dipole or dipolarity — that has been proven in particle physics since 1957 [1,2,3]. Neither have our great national laboratories, etc. made this integration in their official power system thinking. Instead, on energy matters these institutions, organizations, and leaders continue to inappropriately advise the policy makers of the U.S. Government. The government then inappropriately spends the taxpayer's money in the field of energy research, based on that advice.

Consequently, billions of U.S. research dollars are spent annually on an electrical energy science that is archaic and flawed. More billions are spent on energy systems and centralized power grids that are cumbersome, frightfully expensive, and completely vulnerable to modern terrorist attack and natural disasters. These systems are dinosaurs waiting for the terrorist to destroy them.

To "fuel" such power systems, ever more dams, hydrocarbon-burning power plants, pipelines, oil wells, natural gas wells, tanker ships, harbor facilities, refineries, nuclear power plants, and a host of peripheral power systems such as windmills, solar cell arrays, emergency power generators, etc. are required. This great juggernaut also continues to implement a rapacious energy technology which fouls the planet, pollutes the biosphere and destroys much of it. The juggernaut kills off species, is responsible for an uncomfortable and increasing number of human deaths each year from the pollution, and contributes directly to global warming by emitting polluting hydrocarbon combustion products. It places the economy of the United States — so fragilely based on the continued and escalating availability of cheap energy from cheap fuel such as cheap oil and cheap coal — at the mercy of unfriendly states controlling much of the world's supplies of cheap oil. Terrorists are presented with lucrative and strategic soft targets, easily disrupted and destroyed.

The little-recognized basis for such startling technical statements about the powering of electrical circuits has been in particle physics for nearly a half-century. It was evidenced by the award of the Nobel Prize to Lee and Yang in 1957 for their prediction of broken symmetry. The implications of this major discovery — which profoundly impacted all of physics — still have not been incorporated into electrical engineering or the ancient Maxwell-Heaviside-Lorentz electrodynamics taught to electrical engineers.

Consequently, the ubiquitous *vacuum energy* source of all electrical power — for every electrical circuit and electrical power system, small or gigantic — continues to be resoundingly ignored in "energy science and technology", in our universities, and in our leading scientific institutions. The environmental activists, seeking to save the biosphere, have not yet recognized the real problem — the appalling energy advice provided to everyone (including the environmentalists) by the scientific community.

With the above "strong grabbers" to evoke the reader's curiosity and attention, let us explain why such startling and seemingly insane statements are true, how things got that way, and what can be done about it.

Brief History of the Present Classical Maxwell-Heaviside-Lorentz Electrodynamics

In the 1860s James Clerk Maxwell combined electrical fields and magnetic fields into a common model, and launched the present system of classical electrodynamics still being taught today, though in a more limited form. Maxwell's seminal paper [4] was published in 1865, in quaternion-like notation. Quaternion algebra captures many more features and functions of a science which it models, than do either vector algebra or tensor algebra commonly used in electrical engineering.

As an example, with the standard tensor or vector analysis one cannot even "see" the most important functions accomplished by Nikola Tesla in his patented circuits. Quaternion analysis of Tesla's actual patented circuits does show these functions [5]. Consequently, mainstream electrical scientists who use only standard tensor analysis and confidently assume that they understand Tesla's work, are very much mistaken.

In his 1865 paper, Maxwell *specifically lists* his 20 equations and his 20 unknowns. His work was strongly contested, because few of the three dozen electrical scientists on earth at the time were capable in quaternion mathematics.

Before he died in 1879, Maxwell himself had started rewriting his 1873 book for a second edition, with simpler equations.

In the 1880s Oliver Heaviside — a brilliant but self-taught scientist who never attended university — played a major role in converting (reducing) Maxwell's equations to what today is vector algebra, after Maxwell was deceased [6]. Heaviside detested potentials, and stated that they should be "murdered from the theory." The reduction work by Heaviside, Gibbs, and Hertz resulted in the modern four vector equations in some four unknowns. These are taught — along with a further truncation by Lorentz — in every university as "Maxwell's equations". They are in fact Heaviside's equations, further truncated by Lorentz symmetrical regauging [7].

In those early EM days the potentials were thought to be mathematical figments, and all electromagnetic phenomena were considered to be the result of the force fields. Hence any manipulation of the potentials that left the net force fields unchanged, was thought to result in prescribing identically the same systems. Today that is known to be untrue — e.g., in quantum mechanics and quantum electrodynamics, as well as higher group symmetry electrodynamics— but leading classical electrodynamicists still perpetuate the myth.

Both Maxwell's original theory and Heaviside's truncation prescribe two major kinds of electromagnetic systems: (i) those which are "in equilibrium" with their active environment, so they cannot receive *and use* [8] EM energy from it, and (ii) those which are "out of equilibrium" with their active environment, and so can freely receive *and use* [9] EM energy from it.

The first class of thermodynamic systems (those in equilibrium) may be compared to a rowboat floating in a still pond. It has no "net force" upon it, so if we wish the boat to go, we ourselves will have to "row" it, continually putting in force and energy to do work on the boat to force it forward.

The second class of thermodynamic systems (those in disequilibrium) may be compared to a sailboat on the same pond, with a wind blowing. Here we may have to input a little energy to the rudder to "steer" the boat, but the energy and force for the heavy propulsive power is provided freely by the wind. So our boat now "does more work in moving through the water" than the energy that we ourselves input to steer it can do. Simply put, we do not have to *row* the boat, but only *arrange the sails and steer it with the rudder*. The wind puts in the excess energy and force required to propel the boat, so the conservation of energy law is not violated. In simple terms, we *gate and control* the use of more energy than we ourselves have to furnish.

Such a system can even be completely "self-powering", similar to a windmill in the wind or a waterwheel driving a mill for grinding grain. We have to pay to build the windmill or the waterwheel, and to maintain it, but we do not have to input any energy or force to it ourselves, once it's up and running and the wind is blowing or the water is flowing.

The same is true for EM systems, because Maxwell's theory is a purely material fluid theory. Hence anything a fluid system can do, in theory Maxwellian systems can also do because the equations are the same and prescribe analogous functions.

Before Lorentz regauging, the Maxwell-Heaviside equations are still difficult to solve analytically. Numerical methods are often required. This posed a calculation nightmare back in the mid 1800s, before the advent of modern computers and automated calculations. Today, numerical methods can be accommodated much more easily, using computers.

To reduce the difficulty in solving the Maxwell-Heaviside equations and largely eliminate the need for laborious numerical methods, simpler "Maxwellian" equations were sought. Lorentz further reduced the Maxwell-Heaviside equations by "symmetrically regauging" them [10]. This symmetry constrains the modern gauge freedom principle, whereby the potential (and the potential energy) of an EM system can be freely changed at will. In those systems covered by the reduced theory, the potential energy can still be changed. But it can only be changed in such a manner that the two new free fields produced are equal and opposite. Hence the new fields "fight each other to a draw", changing the *internal stress* of the system but doing no *external work* (which requires a net nonzero force field). This has the effect of bottling up any excess EM field energy that might be received by the system from its environment, into a force-free stress potential inside the system The system can be freely energized by the environment to *stress the system,* but it cannot use the free stress potential energy to perform any external work. To perform work, such a system has to have an additional input of energy where a net force field also emerges. In short, the system has to additionally be asymmetrically regauged so as to result in a net force — which means the extra asymmetrical regauging energy has to be input by the system operator or experimenter, since the Lorentz-regauged system itself prohibits the environment from furnishing such "energy with a net field".

In effect, Lorentz modified the equations to select only the far simpler "first class" of Maxwellian systems — those in equilibrium with their external active environment, and thus unable to receive and use any "free energy" from it [11]. This made the resulting equations simpler and much easier to solve analytically. It also inadvertently discarded an entire class of Maxwellian systems — those in disequilibrium with their active environment, able to freely receive excess potential energy and net field energy, develop a net force as a result, and then use that net force to dissipate the excess energy to perform work in an external load.

To ease mathematical solution of the equations, Lorentz *arbitrarily* and unwittingly threw away the electrical windmill and sailboat, and retained only the rowboat. Electrodynamicists and electrical engineers continue to dutifully utilize the Lorentz-regauged subset equations. Consequently, our present electrical power systems — *which are designed and built according to the symmetrized equations* — will not and cannot receive and use EM energy from the many "electrical winds" that can easily be made electromagnetically [12]. By definition, our engineers build only the "first class" of Maxwellian systems, and *never* build a system of the second class. Most no longer even believe that the second class of EM systems exist — *because it does not exist in their archaic 137-year old EM model.* In short, this is the classic case in science where one branch of the scientific community ardently defends an antiquated and imperfect model, even though better models already exist in other branches of science.

Two Kinds of Systems and Two Kinds of Thermodynamics

There are two major kinds of thermodynamics (the science of how energy is dissipated and converted). First, there is the oldest kind, for systems in equilibrium with their environment. This *equilibrium* thermodynamics applies only to systems which do not *receive and use* [13] excess energy from their environment. In short, it applies to the rowboat, not to the windmill in the wind, the sailboat, the waterwheel, the solar cell, the heat pump, etc. For such a system, one must always input more energy to the system than the work we get back out of the system, because some of our input energy is wasted in the system itself (against friction, internal losses, etc.). So its *coefficient of performance* (work out divided by energy we ourselves input) is always less than unity. Or in short, its COP<1.0. In the real world, we can never break even in such a system [14], because the systems we build do have internal losses and inefficiencies.

Hence all our conventional EM power systems exhibit COP<1.0, and have done so for more than a century. Lorentz and our present universities see to it that our engineers design and build only those electrical power systems which *self-enforce* equilibrium conditions, thus obeying the "old" thermodynamics.

The second kind of thermodynamics is for systems not in equilibrium with their active environment. As an example, Ilya Prigogine received a Nobel Prize in 1977 for his contributions to this kind of thermodynamics (of systems far from equilibrium with their active environment). In short, this kind of thermodynamics applies to the windmill, the waterwheel, the sailboat, the solar cell, the heat pump, etc. It also applies to one class of Maxwellian systems, but unfortunately that is precisely the class that Lorentz arbitrarily discarded by his symmetrical regauging. Because this type of system can freely receive *and use* excess energy (it has a net nonzero force field) from the environment, it can output more work than the energy we ourselves input. The excess energy (with appropriate net force field resulting) to do the extra work and power the inefficiencies of the system is furnished by the external environment.

The common home heat pump is a beautiful example of a system far from thermodynamic equilibrium with its environment (the outside atmosphere). By extracting heat energy from the surrounding atmosphere and using it to heat the home, the heat pump under nominal conditions has a theoretical maximum COP = 8.22 [15]. To cool the house, the heat pump reverses to extract heat from the interior air and dissipate it to the external atmosphere. Even with its internal losses, a good home heat pump will produce a COP = 4.0 under reasonable conditions (not too cold outside, etc.). So COP>1.0 systems are well-known — except in electrodynamics, where Lorentz arbitrarily discarded them, where the electrical engineers and power scientists continue to discard them, and where all our electrical power systems for more than a century have been designed and built in accord with Lorentz's deliberate reduction of the theory!

The "Perpetual Motion Machine" Conundrum: An Exercise in Very Bad Logic

The erroneous notion that all EM systems (rather than just Lorentz-regauged EM systems) are "restricted by nature" to COP<1.0 has been ingrained in electrical engineers and electrical scientists for more than a century. Accordingly, an iron dogma has arisen around the issue. Yet it is refuted by every single charge and dipole in the universe [16].

Nonetheless, the great majority of electrical engineers and scientists firmly believe it is physically impossible and against the laws of nature to build an EM system that produces more work output than the energy *we ourselves* input [17]. Proposing such an EM system is considered to be proposing perpetual motion machines that create energy out of nothing. The engineers and scientists consider the proponent of such EM systems a "perpetual motion nut". Yet the very electrodynamics they themselves use and teach already implicitly assumes that every charge and dipole freely creates — and continuously pours out in all directions — EM energy to change the energy density of spacetime and create their associated fields and potentials. In short, by their own "definition" and accusation, every one of them is already a "perpetual motion nut" of the worst kind.

What *is* true is that it is physically impossible to build a system — electromagnetic or otherwise — that outputs more work (conversion of the form of energy) than *all the usable energy that is input to it and made available to it, either by the operator or by the environment, or by both* [18]. As an example, the Bohren experiment [19] — replicable by any good university physics laboratory — always outputs more EM energy than the experimenter inputs (some 18 times as much). As another example, every charge and every dipole in the universe already outputs EM energy continuously, and we do not have to input any energy at all, once either the charge or the dipole is produced and just left alone.

A System Continuously Extracting Energy from the Vacuum Can be Made for a Dollar

Nature readily provides bountiful sources of unending EM energy, free for the taking and using — anywhere, anytime. They are called "charges" and "dipoles" — and often, *source* charges and *source* dipoles. As an example, simply place an electret or charged capacitor on a permanent magnet so that the electric field of the electret or capacitor and the magnetic field of the magnet are at right angles to each other. That silly thing will sit there and pour out EM energy in all directions, at the speed of light, so long as you just *leave it alone and do not destroy it.* One year after you create it, its outpouring of energy will have reached a radius of one light year — out beyond the solar system — in all directions. *It will have changed the energy density of that vast volume of surrounding space of one light-year radius.* And it is still pouring out energy at the same rate, steadily changing the energy density of still more space.

Even the "conventional" electrodynamicists agree that a flow of EM energy is continuously emitted from that arrangement [20]. However, the conventional electrodynamics is resoundingly silent on where that steady outpouring of EM energy *comes from* and *how it is input* to the charge or dipole. There is no *detectable* input of EM energy to the charge or dipole, but there is a detectable and continuous output of energy from it. *Seemingly,* every charge and dipole is creating energy out of nothing, which of course totally destroys the conservation of energy law if true [21]. Since classical electrodynamicists have not been able to solve it, their textbooks have remained very silent on this fundamental problem and its implications. *By their resounding silence, classical electrodynamicists implicitly assume that every source charge and source dipole in the universe is a perpetual motion machine, freely and continuously creating energy out of nothing.*

Either one rejects the energy conservation law entirely, or one explains the source charge problem. There is no middle position, because the source charge is real and it ubiquitously exists. And it ubiquitously pours out that energy.

So we point the finger right back at the self-appointed critics and ask in their own terminology, "Who are the real perpetual motion nuts here? You cannot logically consider and implicitly accept every charge and dipole as a perpetual motion machine, freely creating energy out of nothing at all, and then protest that there can be no such thing as a COP>1.0 EM system!"

All the EM energy in any EM circuit or device comes from those same source charges and source dipoles. *If one cannot explain where and how those charges obtain the energy to keep pouring it out, then one knows absolutely nothing about what truly powers every electrical circuit. Since the energy being received is nonobservable, it must be received in some peculiar and normally unusable form. The charges must then transduce the received energy into usable and observable form, and re-emit it in that new form, so that the circuit can catch some of it and be "energized".*

The source charge problem focuses one's attention on the real problem. Either we must discard energy conservation altogether, or we must admit that every charge and dipole is already a COP>1.0 Maxwellian EM system, freely changing the form of some unusual received energy. It is continuously doing "free" work, since work is the changing of form of energy, and every charge and dipole is freely receiving virtual EM energy and changing its form to observable EM energy, continuously.

The quandary of the source charge and its continuous outpouring of real EM energy has been called "the most difficult problem in classical and quantum mechanics" [22]. Until 2000 there did not appear any classical solution to that long-vexing problem of the association of the fields and potentials — and their energy — reaching across all space, with the source charge that produces them. But the basis for the solution was already present in particle physics since the discovery of broken symmetry in 1957. In 2000 the present author proposed a formal solution [23] consistent with quantum field theory [24], particle physics [25], quantum electrodynamics [26], and re-interpretation 12 of Whittaker's biwave decomposition of the scalar potential [27]. We used the term "giant negentropy" since the charge and dipole continuously and freely absorb, cohere, organize, and re-emit energy from the vacuum.

Since every charge in every circuit is already continuously negentropic, then our building of entropic circuits using these negentropic charges must involve some characteristic of circuit design where we kill the negentropy. We will explain that aspect shortly.

Our universities should — but do not — focus on the main problem: *How does one then intercept, divert, and collect some of that freely flowing EM energy so easily evoked by every charge and dipolarity, and use it to freely power loads, without disarranging and destroying the actual "power source"?* No university seems to be working on that problem, nor is the Department of Energy, nor are any of the great national laboratories. Nor are any of the great scientific associations.

Yet *that single problem is the only fundamental electrical power problem.* All the rest of the "recognized power problems" are just so much psychological displacement activity so as not to have to disturb the comfortable Lorentz-regauged and crippled classical electrodynamics.

There is no problem at all in obtaining great rivers and gushers of EM energy from the ubiquitous vacuum — cheaply, easily and enduringly. Every charge and every dipole already does that. There is no problem in producing the free "electromagnetic energy winds" needed to power even the greatest loads, any place in the universe, any time. Instead, the only problem is in building a proper "electrical windmill" to divert, collect, and use some energy from such a steady electrical wind — *without destroying the broken symmetry source of the wind (the source dipole)* — and using (dissipating) that collected energy to power our desired load.

All our present closed-current loop circuits are designed [28] to use half their freely collected energy to destroy the source dipole (in the generator or battery). The other half of their collected energy is used to power the external circuit's losses and loads. So more of the freely collected EM energy is used to destroy the "wind" source than is used to power the load. We then have to continually input at least as much energy to restore the source (the dipole) as was used to destroy it. So we must continually input more energy to restore the dipole — that the engineers unwittingly design and build the circuits to deliberately destroy — than the circuit provides to power the loads. All our engineers design and build electrical power systems that destroy their free electrical wind sources faster than they power their loads. Such inane power systems obviously provide COP<1.0, because of the deliberate circuit design used by the engineers.

It is akin to building a magnificent but flawed sailboat, which — once it starts to move in the wind — rapidly and continually lowers its own sails faster than the wind can propel the boat. To keep the boat going, one thus has to continually pay to keep raising the sails that the boat itself keeps lowering. Our electrical engineers and electrical power scientists are busily engaged in perpetuating an analogous electrical power technology, and our scientific community is busily assuring us that they are practicing "the most advanced electrical science".

We actually pay the power company to engage in a giant sumo wrestling match inside its generators, *and deliberately lose.*

The Modern Vacuum: Empty Space Has Unlimited EM Energy

"Vacuum" is what we usually think of as the empty space left after all the air is removed. However, in modern physics it is well-known that space is not such an "emptiness" at all. Instead, it is filled with energetic particles that appear and disappear with extraordinary speed. Hence "the vacuum" in physics is more like a seething cauldron, boiling fiercely [29]. The energy density of this boiling is so great that it literally boggles one's mind [30]. A bit of "empty space" the size of a sugar cube contains enough "seething electromagnetic energy" to power all the electrical loads on earth for millions of years. Obviously, if we can extract just a tiny fraction of this energy in electromagnetic form, we can use it to power our loads for free or nearly so. We need only power the "gating" or "switching" of the flowing energy, and thus cause a great deal of work to be performed, even though we ourselves input little or no energy.

Can the EM Energy of the Vacuum be "Tapped" and Extracted for our Use?

It's widely believed that it is impractical to try to extract very much usable energy from the seething, highly energetic vacuum. Oh yes, the tiny little Lamb shift [31] of the orbit of one electron in a certain atom is admittedly due to vacuum energy interaction. Well, although the Lamb shift is a very small effect of a single electron in a single atom, the *energy density* of that interaction is greater than the surface energy density of the sun [32]!

And, oh yes, the Casimir effect [33] — a vacuum-induced attraction of two conducting plates when separated but close — also is due to vacuum energy interaction, but it too is a very small effect and not practical to consider [34].

Then the assertion is often made — particularly by electrical power engineers, who normally are not well-acquainted with modern physics — that thermodynamics prohibits extracting and using vacuum energy. That is quite untrue, as we have explained, and as Cole and Puthoff have rigorously demonstrated [35]. *Thermodynamics does not explain how to do it practically, but it does permit it technically.*

The principle of superposition of charge and the field effects of charges, ubiquitously assumed and verified in countless classical experiments, also tells us a magnificent thing: *The source charge production of its associated fields and potentials is additive and coherent as the source charge is increased . Within reason, doubling or tripling the amount of energy-collecting and transducing charged particles (source charges) at a given point will give double or triple the resulting potentials and fields (gathered EM energy and local EM energy winds) produced by that source charge collection, and so on.*

So the proven broken symmetry of opposite charge collections — such as on the ends of a dipole or dipolarity — need not be a small effect at all! We can use as many unit charges as we wish in each one of those "separated opposite charges" in a dipole, so long as we hold the charges apart with a proper restraining dielectric. In a normal electrical circuit (as between the terminals of a generator or battery), we can have many positive and negative charges involved, so that the EM energy extracted from the vacuum can be very powerful. In other words, the dipole can readily be made essentially as powerful as we need for our usual specific purposes. We can easily produce a megawatt or even 1,000 megawatts of energy flow per second, by making a bigger dipole consisting of more separated charges.

All we have to do is pay to make the dipole once. Then leave it alone and do not allow it to be destroyed by the inane way that engineers normally build their circuits and power their loads.

The latter problem — which is the only real problem preventing the engineers from easily producing all the cheap, clean electrical power they wish, without burning fuel, consuming fuel rods, building pipelines, building dams, etc. — is not being furiously funded and worked on as a national Manhattan Project. Indeed, the very notion continues to be castigated and ridiculed by the "defenders of the faith". And that is perhaps the greatest scientific travesty of our times. It simply reveals the "soft underbelly" of our present energy science, which does not even include what powers its circuits.

Can an Electrical Power System Output More Energy than We Input to It?

Hopefully, by now the reader will exclaim, "Of course! Just pay to make a source dipole." Prepare a suitable interception and collection external circuit which will dissipate its collected energy in a load, without destroying the source dipole faster than it powers the load. That's all there is to it. The only remaining problem is in that single phrase, "without destroying the source dipole faster than it powers the load".

Since that is the problem and the only problem preventing a total solution to the energy crisis forever, every university, every power company, the National

Science Foundation, the National Academy of Sciences, the Department of Energy, and every major research lab should be working on it. But of course none of them is working on it, interested in it, or funding it.

The problem calls for "out-of-the-box thinking", to use a prevailing buzzword.. Simply consider a charged capacitor or a permanent magnet. That beast is freely extracting and pouring out real EM energy, from the active vacuum, continuously. So how do we intercept some of the energy flow, convert the intercepted energy to, say, oscillating EM energy, and then use that energy to power a load — all without destroying the magnet or killing the charge in the capacitor?

This is the problem, and work on it by our sharp young graduate students and post-docs is something which the National Science Foundation and the National Academy of Sciences should fund as the highest priority. They do not even recognize the fundamental problem, much less fund work on it.

These agencies' lack of recognition of this, the primary electrical energy problem, is a sad commentary on our present scientific acumen.

Why Don't Present EM Systems Power Themselves?

By now the reader hopefully will be able to see that our systems don't power themselves because our engineers specifically design them so they *cannot*. The mindset of the scientists and engineers themselves is the greatest foe to cheap clean electrical energy, a clean biosphere and a decentralized power system using no fuel. The closed current loop circuit for power systems is considered to be hallowed, as if Moses had brought it down from the mountain with him, carved on those stone tablets. It isn't and he didn't.

To quote Nikola Tesla, who made practical generators and motors possible with his discovery of the rotating magnetic field: *"Today's scientists have substituted mathematics for experiments and they wander off through equation after equation and eventually build a structure which has no relation to reality."* Understand, a good model is essential to technology. But if the model is inadequate, then it should either be corrected or a new model should be substituted that is adequate. However, our electrical power engineers and most of the scientific community have developed a mindset where the closed current loop circuit, Lorentz symmetrical regauging and COP<1.0 electrical circuits are considered to be iron laws of nature. They are not such at all, but merely the self-limited and self-enforced contrivances of human beings.

We should always remember one succinct statement by Evans [36]: *"No theory can falsify a successful and repeatable experiment, but a single successful and repeatable experiment can falsify any theory."*

We borrow a quote from Tesla on another subject [37]and hope that it comes true with respect to the present dogma that only COP<1.0 electrical power systems are possible: *"... in a short time it will be recognized as one of the most remarkable and inexplicable aberrations of the scientific mind which has ever been recorded in history."*

What Is Significant About More Modern Kinds of Electrodynamics?

The present Lorentz-regauged Maxwell-Heaviside theory effectively assumes an inert vacuum — falsified for half a century by quantum mechanics and particle physics — and no curvatures of local spacetime — falsified for nearly a century by general relativity. Since the active vacuum and the local curvature of spacetime — and their dynamics — are the "active external environment" for an electrical power system, then *present power engineers and classical electrodynamicists assume away any and all net interactions of the system with its external active environment.* That is how the COP>1.0 Maxwellian systems are arbitrarily discarded, and only electrical power systems that self-enforce COP<1.0 are designed, built and permitted. To then drum into the students' heads that this travesty represents a "law of nature" is ludicrous. It has become *disinformation, propaganda* and *dogma,* and it violates the very spirit of scientific inquiry and scientific reasoning.

The proper scientific frame of mind is skepticism tempered by open-mindedness. If one is skeptical and dogmatic, one is not scientific, regardless of one's pedigree. If one is open-minded and not skeptical, one is scientifically naïve.

Meanwhile, much more modern systems of electrodynamics have been developed in particle physics and elsewhere. In modern theory, the models are based very powerfully on the theory of groups. Groups have symmetries, designated by such expressions as U(1), SU(2), O(3), etc. The higher the group symmetry utilized by an algebra, the greater range of physical phenomena to which the algebra can apply and can model.

A particularly good higher group symmetry electrodynamics, in this author's opinion, is the O(3) electrodynamics founded by Evans and Vigier and further expounded by Evans [38]. Evans has shown that O(3) electrodynamics is a part of the Sachs unified field theory electrodynamics [39]. Thus O(3) electrodynamics can be used not only for modeling "normal" electrodynamic but also for modeling "exotic" unified field theory. Further, it can be used for engineering, so it permits the development of a drastically extended electromagnetic technology which can eventually engineer many new phenomena [40], including antigravitational effects [41]. At least one highly successful antigravity experiment was performed by Sweet, in an experiment designed by the present author [42]. The weight of an object was steadily reduced by 90%, on the laboratory bench.

Why Don't the Electrical Engineering Departments Teach a More Modern Science?

Here the reader's assessment is as good as anyone else's. After more than a century, the standard Lorentz-regauged Maxwell-Heaviside dogma has become so ingrained in electric engineering thought that everyone accepts these equations without question. To use an appropriate phrase from Einstein, it is as if we "imbibed it with our mother's milk." Behind the scenes, of course, there are also very powerful financial interests — which Winston Churchill just loosely referred to as the "High Cabal" — with income of some trillions of dollars per year at issue. These groups do not wish the present energy science and technology changed into something dramatic and cheap such as "fuel-free" electrical power systems extracting their energy from the vacuum. It has already been decided that the path of so-called "progress" in power systems for the future will consist of fuel cells and the more difficult hydrogen technology. This will keep the environmentalists off the cartels' backs, and allow them to show "progress" being made to clean up the biosphere.

The bottom line is that it's still all about money and power, and "who's going to be the big monkey". Primate dominance is still one of the primary motivations in humankind, including in scientists and the scientific community.

The terrible bias in electromagnetic science toward that flawed old energy model is guaranteed by the "scientific momentum" and inertia that the dogma has built up over more than a century. Science does not change its cherished dogmas by sweet reason, but by a vicious and sustained cur dog fight — and even then only after the "old dogs" die off and get out of the way. As Max Planck put it [43]:

"An important scientific innovation rarely makes its way by gradually winning over and converting its opponents: it rarely happens that Saul becomes Paul. What does happen is that its opponents gradually die out, and that the growing generation is familiarized with the ideas from the beginning."

Arthur C. Clarke said it this way [44]:

"If they [quantum fluctuations of vacuum] can be [tapped], the impact upon our civilization will be incalculable. Oil, coal, nuclear, hydropower, would become obsolete — and so would many of our worries about environmental pollution." "Don't sell your oil shares yet — but don't be surprised if the world again witnesses the four stages of response to any new and revolutionary development: 1. It's crazy! 2. It may be possible -- so what? 3. I said it was a good idea all along. 4. I thought of it first."

It's a part of human nature to strongly resist new ideas, and particularly *upsetting* ideas. Even Sigmund Freud, the founder of modern psychology, pointed it out in these words:

"There are three steps in the history of a great discovery. First, its opponents say that the discoverer is crazy; later that he is sane but that his discovery is of no real importance; and last, that the discovery is important but everybody has known it right along."

Just now, COP>1.0 energy science and notions are still largely in that first stage prescribed by Freud, just a little bit into the second stage, and struggling to get on into and through the second stage and into the third.

A key problem is that all science must be patronized (funded). Someone must pay for the labs and the scientists, the equipment, the operations, etc. The scientific community is organized to strongly *control* that funding, and *control what it can be spent for*.

When the research money is strongly controlled, scientific research itself is strongly controlled. So the real reason that this type of research does not happen is that managerial and control scientists are rewarded — often handsomely — for adhering to the status quo, and doing only "incremental improvements" rather than gigantic leaps. Scientists, graduate students and post-docs are in fact destroyed or severely punished if they "wander off" into that old "perpetual motion machine nonsense" called COP>1.0 EM systems. The scientific "system" is set up that way, very much like a cartel, and it is self-enforcing and self-replicating. *Scientists are allowed to work on only what they can attract funding for*. They are only funded to work — mostly — on comfortable and already established things.

So ultimately the blame lies directly with the leaders of the scientific community, such as the National Science Foundation and the National Academy of Sciences, and the manner in which they advise the U.S. government to tailor its energy research money. Also, as a result the U.S. Patent Office still is very shy about granting any kind of patent that mentions or implies COP>1.0 EM systems. Such is simply assumed to be against the laws of nature and involving "perpetual motion nonsense", *a priori*.

Doesn't Present Circuit Theory Accurately Predict All Circuit Behavior?

Not on your life! Circuits do continue to exhibit anomalous behavior, and behavior that is not understood. Even many of the basic effects are not well-understood. For example, there are about 200 known magnetic effects, and only about 100 are well-understood. The level of understanding of the remainder varies from "somewhat" to "not at all".

For nonlinear circuits, the phenomenology can get very anomalous. E.g., quoting Ogorzalek [45]:

> *"All real systems are nonlinear in nature. This simple observation is true also for electrical and electronic circuits even though many of them are designed to perform linear transformations on signals. ...in many cases the designed circuit, when implemented, performs in a very unexpected way, totally different from that for which it was designed. In most cases, engineers do not care about the origins and mechanisms of the malfunction; for them a circuit which does not perform as desired is of no use and has to rejected or redesigned...."*
>
> *"...electrical and electronic circuits constitute a group of real physical systems in which observations and measurements are relatively easy to make. ..Such an 'experimental comfort' enabled thorough studies confirming the existence of strange unexpected behavior in almost every type of electronic circuit – oscillators, filters, instrumentation circuits, power supplies, PLLs, electric machines, microwave circuits, electro-optic systems, etc. The main problem remains in interpreting experimental data."*

As another example, the "open path" discovered and utilized by Gabriel Kron [46], one of our great electrical scientists, is still unknown today. Kron was never allowed to reveal the full exposé of his open path, where he could produce current between any two parts of a circuit, even though not connected. Lynn and Russell [47] pointed it out this way:

> *"Kron has never published details of his method of making the polyhedron self-organizing, although his published results show that in this state it has some remarkable properties, associated with harmonic integrals on multiply connected spaces."*

For the uninitiated, in a multiply-connected space some energy can be in two or more places at the same time. Contrary to our intuition, some parts of physics really do work this way. Obviously, if one inputs 100 joules to one point in space, and it remains there yet also appears simultaneously in another point in the observer's singly-connected space, this appears to be "creation of energy" to the observer. It is not. Instead, it is a broken symmetry in energy conservation in singly-connected space, resulting in a higher symmetry in multiply-connected space. Conservation of energy laws need not take such a simple form as our naïve intuition would have us believe at first blush, or that almost all the textbooks proclaim,

Energy in multiply connected spaces is often referred to as a *quantum potential*. Bohm's great hidden variable theory of quantum mechanics [48] uses the quantum potential, and it poses things eventually engineerable that are presently deemed impossible. His theory gives all the proper predictions and answers to the current problems and experiments, and it is equally as valid as the present much more limited Bohr interpretation. Indeed, Bohm's theory seems to gain ground with every passing year [49].

Why Don't Scientists and Engineers Develop Power Systems That Extract Our Electrical Energy Freely from the Vacuum?

Scientists do not research electrical energy from the vacuum because the scientific establishment will not allow such research to be funded. They cannot teach it, for it would not be allowed. Further, if they try to do it on their own, they get no funding, their papers are denied publication and their peers destroy them. Eventually they will be lucky to be employed in a fast food restaurant.

The system has become "self-policing" of its own cherished dogma. The reviewers definitely deny an applicant funding if the project submitted would be an embarrassment to their own self-interest. E.g., Charles McCutchen, a physicist with the NIH, puts it this way: "*[Peers on the panel reviewing a grant applicant] '...profit by his success in drawing money into their collective field, and by his failure to do revolutionary research that would lower their own ranking in the profession. It is in their interest to approve expensive, pedestrian proposals.'*"

Well, surely the scientific journals will readily publish innovative new things, will they not? Again, not so. To quote George [50]:

"…I suggest that most revolutions in science have taken place outside the lofty arena of the refereed journals, and with good reason. The philosophy by which these journals govern themselves virtually precludes publication of ideas that challenge an existing consensus."

In that respect, the publication of two papers (40) by the Alpha Foundation's Institute for Advanced Study (AIAS) in Foundations of Physics Letters, on a COP>1.0 EM system, was nothing short of miraculous. With the publication of several other AIAS papers dealing with energy from the vacuum [51], a small salient in the solid opposition of the journals appears to have at last been opened. But there is a very, very long way to go before such papers can be published readily, even when meticulously prepared and scientifically impeccable.

Why does not one paper do it? Why is such a volume of papers needed? As Landauer puts it [52]:

> *"We used to be able to say things once; if the message was reasonable, it had a good chance of becoming a permanent part of the structure of the field. Today, a single publication is lost; if we say it only once, it will be presumed that we have changed our mind, and we therefore must publish repeatedly. This further fuels the large publication volume that requires us to repeat."*

Actually this is one somewhat unfortunate aspect of science's success and growth. Quoting Price, nearly four decades ago [53]:

> *"Using any reasonable definition of a scientist, we can say that between 80 and 90 per cent of all the scientists that have ever lived are alive now. Now depending on what one measures and how, the crude size of science in manpower or publications tends to double within a period of 10 to 15 years."*

In short, because of the sheer bulk of the scientific literature and its ever-expanding volume, the subject of a paper can just be lost in the wilderness, unless it is extensively published in many papers, and in multiple journals.

In the field of permissible COP>1.0 power systems, still struggling to be born, the resistance assumes a more sinister, Taliban-like character. The proponent is assaulted with a viciousness and a tirade seldom equaled in the annals of scientific history. Few can survive such an onslaught. Here is a direct example of the type of vicious assault savagely made upon a reputable cold fusion scientist — at the time, cold fusion had achieved more than 600 successful experiments by multiple scientists in multiple labs in several nations:

"How stupid do you think we are? My assessment of you and your colleagues is that you are complete frauds or totally mad. There is no known physical principle that would support the kind of results that you claim your technology can accomplish, nor is there any credible argument why there should be such a principle.

In science, of course, one is supposed to believe and accept the experimental results, if they are replicable. It does not matter whether a "known physical principle" is available for the experiment, or whether a "credible argument" for it even exists. Either we follow that iron dictum of the scientific method, or we do away with scientific method altogether and practice "truth by authority and decree". Further, calling more than 100 scientists, who have performed successful experiments and replications, frauds or madmen, is totally an ad hominem attack and has no place in science of scientific method.

Any historian of science can cite many dozens of incidents where ruthless suppression of new and important scientific discoveries has been done by the scientific community itself. Science has had *and continues to have* a dark history in that respect.

How Will "Electrical Energy from the Vacuum" Science Eventually Get Started?

It will get started when one or more of the inventors and researchers somehow succeeds in getting sufficient funding and scientific support (he will have to pay the scientists himself) to develop a simple, rugged, easily replicable COP>1.0 electrical power system and place it on the world market. Even then, he will have to do it outside the United States, in all probability. Otherwise, the U.S. scientific establishment will destroy him, by fair means or foul. And we do mean *foul*. Jim Watson demonstrated an 8 kilowatt practical self-powering generator system at a national conference and then apparently received an "offer he could not refuse". He and his family simply dropped out of sight and severed all further contact with his friends, including this author. He and his family are still alive, fortunately, but remaining very, very quiet. There are no Watson systems on the open market and there are not going to be.

Steve Marinov was killed in a foreign country with a longitudinal EM wave "shooter", and his body was thrown from the top of a building to suggest a suicide. The body was left there for an extended period by the authorities, who knew it was emitting longitudinal EM radiation for awhile. Where the body lay on the concrete, the cement glowed. There is only one weapon on earth that will kill a body in such fashion: a longitudinal EM wave shooter. At least 10 nations of the world have developed longitudinal EM wave weapons. These are almost certainly the weapons to which Secretary of Defense Cohen referred in 1997 as follows [54]:

> *"Others [terrorists] are engaging even in an eco-type of terrorism whereby they can alter the climate, set off earthquakes, volcanoes remotely through the use of electromagnetic waves... So there are plenty of ingenious minds out there that are at work finding ways in which they can wreak terror upon other nations... It's real, and that's the reason why we have to intensify our [counterterrorism] efforts."*

Sweet was fired at from about 300 yards by a would-be assassin, using a silenced rifle. Being old, he stumbled and fell on the steps just as the assassin pulled the trigger. The bullet snapped right by his ear, where his head had just been. Thereafter, Sweet was always deeply paranoid about taking the unit outside his own apartment or continuing to develop it. Again, I personally worked with Sweet for some years.

The present author has had his own share of such shenanigans, close calls and suppressions, more than a dozen times but does not wish to publicize the incidents. As a single personal example, Kawai's patented motor [55] will produce COP>1.0 if his process is applied to a high-efficiency magnetic motor to start with (such as one from Hitachi with 70% to 80% efficiency). We explained the Kawai engine's operation, placed it on the Internet, and Kawai and party came to Huntsville to see us. At Kawai's urging, we negotiated an agreement that we would manufacture and market Kawai's systems worldwide; he already had built a closed-loop, self-powering system in Japan. He would fund the entire project. Our agreement was verbally reached on a Thursday afternoon, late. That night a jet arrived posthaste from Los Angeles, with a Yakuza on board. The next morning Kawai and his party were in fear and trembling, and the Yakuza was in total control. Kawai no longer controlled his own company, his invention or his own fate. Needless to say, the Yakuza coldly cancelled the agreement, point blank. This happened in front of my associates and I, so there are multiple witnesses. The Yakuza and party quickly packed up the two Kawai engines in our possession and departed. No Kawai engine will ever be permitted on the world market. Several other Japanese COP>1.0 electrical power systems have also been suppressed by the Yakuza [56]. Many such incidents — including murder — have occurred over the last decades, right here in the United States. Others will happen.

The single thing that is likely to finally force COP>1.0 systems out there is the Internet. For the first time, researchers can exchange information worldwide and on a vast scale. My own website, www.cheniere.org, is an example of a tired old dog, trying to get out important information he has discovered — or believes he has discovered — and make it available freely to students, professors, scientists, and researchers that are interested. Jean-Louis Naudin's website, http://members.aol.com/jnaudin509/, is another example. There are many others.

Of course, there is a great deal of *dis*information on the web as well as information. The reader still has to sift it for himself. In my own case, I extensively cite the references I use, and the interested party can check them himself to see if they say what I said they did. In short, the idea is to educate, not dogmatize. It is also intended to give the interested young researcher what I have uncovered in 30 years of very hard work, so he does not have to rediscover all the wheels again, so to speak. He can simply start from where

I am and go much further. It is very heartening that many other researchers are doing the same thing and sharing what they have uncovered. In my own case, I have also labored very hard to lay a sound basis for a legitimate theory of permissible COP>1.0 EM systems. That effort will culminate with the publication of my book late this year [57].

As more researchers succeed, eventually this thing will finally acquire enough momentum to "get born". When the scientific community finally turns the sharp young graduate students and post-docs loose on it and funds them, it will simply explode into science and technology like a skyrocket. Then, of course, it will be just as Planck and Freud stated. Suddenly everyone will have not only *known* it all along, but *they did it first*.

But it *will* get born. If it doesn't, the "dirt economics" in poor nations with fanatical leaders — and the increasing struggle for control of cheap oil and its use as a weapon —will generate wars and world terrorism. The terrorists will see to it that humanity destroys or nearly destroys itself during the first half of this century, and perhaps even by the end of this first decade. It is not accidental that President George W. Bush has declared war on terrorism.

Every nation must soon get cheap clean electrical energy from the vacuum and develop self-powering electrical power systems quickly or no nation will ever be safe again. Radicals and terrorists will see to it, if we lose the energy war.

So COP>1.0 electrical power systems have become both a moral imperative and a national imperative, in my opinion. Either we develop them and establish them worldwide or we perish, as simple as that. If enough of us determine to do it, and if we hang together and get a little funding — and if we have a little luck— hopefully it will be accomplished.

(Endnotes)

1. Lee and Yang strongly predicted broken symmetry in 1956; see T. D. Lee, "Question of Parity Conservation in Weak Interactions," Physical Review, 104(1), Oct. 1, 1956, p. 254-259; — and C. N. Yang, "Remarks on Possible Noninvariance under Time Reversal and Charge Conjugation," Physical Review, 106(2), 1957, p. 340-345.
2. Wu *et al*. experimentally proved broken symmetry in 1957; see C. S. Wu, E. Ambler, R. W. Hayward, D. D. Hoppes and R. P. Hudson, "Experimental Test of Parity Conservation in Beta Decay," Physical Review, Vol. 105, 1957, p. 1413.
3. So revolutionary to physics was this discovery of broken symmetry that the Nobel Prize was awarded to Lee and Yang in the very same year that Wu experimentally proved it. For Lee's Nobel acceptance speech, see T. D. Lee, "Weak Interactions and Nonconservation of Parity," Nobel Lecture, Dec. 11, 1957. In T. D. Lee, Selected Papers, Gerald Feinberg, Ed., Birkhauser, Boston, 1986, Vol. 1, p. 32-44.
4. James Clerk Maxwell, "A Dynamical Theory of the Electromagnetic Field," Royal Society Transactions, Vol. CLV, 1865, p 459.
5. See Terence W. Barrett, "Tesla's Nonlinear Oscillator-Shuttle-Circuit (OSC) Theory," Annales de la Fondation Louis de Broglie, 16(1), 1991, p. 23-41. Tesla was able to shuttle the potential (and the energy) around in his circuits at will. Barrett, one of the pioneers of ultrawideband radar, extended Tesla's method and obtained two U.S. patents for use in signaling science: See T. W. Barrett, "Active Signalling Systems," U.S. Patent No. 5,486,833, issued Jan. 23, 1996; — "Oscillator-Shuttle-Circuit (OSC) Networks for Conditioning Energy in Higher-Order Symmetry Algebraic Topological Forms and RF Phase Conjugation," U.S. Patent No. 5,493,691, issued Feb. 20, 1996.
6. When Maxwell died, he himself was engaged in converting his own quaternion-like theory into the much simpler vector theory, in a desperate attempt to reduce the objections to it. See Maxwell, James Clerk, A Treatise on Electricity and Magnetism, Oxford University Press, Oxford, 1873, Second Edition 1881 (Maxwell was already dead), Third Edition, Volumes 1 and 2, 1891. Foreword to the second edition was by Niven, who finished the work as Maxwell had dramatically rewritten the first nine chapters, much new matter added and the former contents rearranged and simplified. Maxwell died before finishing the rest of the second edition. The rest of the second edition is therefore largely a reprint from the first edition. The third edition edited by J. J. Thomson was published in 1892, by Oxford University Press, and later was published unabridged by Dover Publications, New York, 1954. J. J. Thomson finished the publication of the third edition, and wrote a "Supplementary Volume" with his notes. A summary of Maxwell's modified equations are given in Vol. II, Chapter IX of the third edition. However,

Maxwell had gone (in his second edition) to some pains to reduce the quaternion expressions himself, and therefore to not require the students to know the more difficult calculus of quaternions (so stated on p. 257).

7. Actually the first "symmetrizing" truncation of Maxwell's equations was by Ludwig van Lorenz, shortly after Maxwell's 1865 definitive paper was published. See Ludvig Valentin Lorenz, "On the identity of the vibrations of light with electrical currents," Philosophical Magazine, Vol. 34, 1867, p. 287-301. In this paper L. V. Lorenz gave essentially what today is called the Lorentz symmetrical regauging. However, when H.A. Lorentz (a different individual with a similar name), then perhaps the greatest electrical scientist, regauged the Maxwell-Heaviside equations in the 1880s, the H.A. Lorentz regauging was adopted because of Lorentz's prestige.
8. The symmetrized systems can and do receive the excess energy only in the form of a net force-free stress potential (simply examine the regauging equations and the regauging condition). With no net force available to dissipate the stress potential energy, no net work can be done by the excess stress energy received. That is precisely what the Lorentz condition means, in simpler terms. The closed current loop circuit automatically self-applies Lorentz regauging, by making the back-emf precisely equal to the forward emf, so that the collected energy becomes stress energy with no net force fields self-generated by the system itself. In that case, the operator must continue to input some form of net work and net force field, to which the system will respond by "using it" to produce the back emf. In that use, the system is able to power a load to that extent only. The known absence of Newton's third law from electrodynamic fields already assures us that back-emf equal to forward emf is not a law of nature in electrodynamics. Yet mainstream electrical engineering persists in assuming that it is.
9. Disequilibrium systems produce net force fields along with receipt of the excess energy — something we have called "asymmetrical regauging". This violates the Lorentz condition, and the system can then utilize this net force to dissipate the net energy received, thus performing some "free work". This is no more mysterious than a windmill receiving energy and net force from the environment, and using that force to dissipate the energy and do free work. To imply that it is somehow a violation of natural law or a violation of energy conservation is .
10. To see how this is done, see J. D. Jackson, Classical Electrodynamics, Second Edition, Wiley, New York, 1975, p. 219-221; 811-812.
11. Again, a system in equilibrium can and does receive excess energy from the environment, but it also immediately returns the energy. The "force in" and "force out" balance to a net zero, so input of additional energy just results in the production of excess system stress, but no free work in an external load.

12. Any dipole (such as a permanent magnet) or charge (such as an electron) pours out EM energy in all directions at the speed of light — and hence produces a continuous "free electromagnetic wind". If we leave it intact, the dipole or the charge will pour out such energy indefinitely. The dipoles and charges in the original matter of the universe have been doing it for some 15 billion years, and they are still doing it. So the process does not "run down". Every charge and dipole in the universe exhibits *giant negentropy*, producing all that EM energy and the EM fields and potentials in nature. Yet for more than a century we have designed and built only entropic systems. We have steadily despoiled the biosphere in the process.
13. We stress this point most strongly. Such a system may freely receive excess energy from its environment, but it self-enforces equilibrium by applying some form of Newton's third law, to negate any net force fields resulting from the excess energy. Consequently, this system can change its stress energy, but not its usable net field energy to do external work. It can only change the equilibrium condition, with a change of stress potential as it changes the conditions of equilibrium.
14. Superconductive *sections* in a system are often touted as COP = 1.0 systems, by implication. However, when the cryogenics overhead one must pay is counted, they are grossly underunity.
15. David Halliday and Robert Resnick, Fundamentals of Physics, Third Edition Extended, Wiley, New York, 1988, Vol. 1, p. 518.
16. E.g., see M.W. Evans, P.K. Anastasovski, T. E. Bearden et al., "Explanation of the Motionless Electromagnetic Generator with O(3) Electrodynamics," Foundations of Physics Letters, 14(1), Feb. 2001, p. 87-94; — "Explanation of the Motionless Electromagnetic Generator by Sachs's Theory of Electrodynamics," Foundations of Physics Letters, 14(4), 2001, p. 387-393.. The tired old "perpetual motion nonsense" charge was strongly raised in the vigorous refereeing of these two AIAS papers. To refute the charge raised against the latter paper, I prepared a vigorous rebuttal, "On Permissible COP>1.0 Maxwellian Systems: A Reply to the Board Member," and submitted it to the referees and editors. Based on the paper, the referees and editors overruled the objections and published the second paper because every charge and every dipole in the universe already clearly demonstrates a COP>1.0 Maxwellian system. We challenged the classical scientists to provide a solution to the source charge problem in their Lorentz-regauged theory, then showed the solution which can only appear in a higher symmetry electrodynamics — such as the O(3) EM or the Sachs EM used in the two papers. The old EM assumes an inert local vacuum and a locally flat spacetime, both assumptions being long since refuted in modern physics. And every charge and dipole demonstrates the falsity of those two assumptions, as is easily demonstrated on the lab bench.

17. Except the classical electrodynamicists who have not resolved their own source charge problem, no one advocates that an EM system can do more work in a load than the *total amount* of usable energy that is input to it! Instead, one advocates that it can do more work than the *operator* inputs; the remaining energy is freely input from the active external environment due to the disequilibrium (the broken symmetry) condition. The broken symmetry also specifically implies violation of the Lorentz symmetry condition, a priori.

18. The reason is extraordinarily simple, once one makes clear definitions of terms. *Work* rigorously is *a change the form of energy*. A priori, an inert system cannot "convert the form of energy" it has not received and collected. But it can receive additional energy from the environment and convert more than the *operator himself* inputs, in which case the environment freely inputs the remainder of the input energy being converted in form!

19. Craig F. Bohren, "How can a particle absorb more than the light incident on it?" American Journal of Physics, 51(4), Apr. 1983, p. 323-327. Under nonlinear conditions, a particle can absorb more energy than is in the light incident on it. Metallic particles at ultraviolet frequencies are one class of such particles and insulating particles at infrared frequencies are another. See also H. Paul and R. Fischer, {Comment on "How can a particle absorb more than the light incident on it?'}," Am. J. Phys., 51(4), Apr. 1983, p. 327. The Bohren experiment is repeatable and produces COP = 18.

20. As an example, quoting Jed Z. Buchwald, From Maxwell to Microphysics, University of Chicago Press, Chicago and London, 1985, p. 44: *"[Poynting's result] implies that a charged capacitor in a constant magnetic field which is not parallel to the electric field is the seat of energy flows even though all macroscopic phenomena are static."*

21. The key is in that word "detectable" — or as the physicist would say, "observable". The experiment rigorously demonstrates that, if the law of energy conservation is valid, then the charge and the dipole must be continuously receiving energy (from its external environment) in nonobservable (virtual) form and integrating it into observable form. And so it is, as has been known in particle physics for 45 years. In the case of the dipole, it is due to the known broken symmetry of the opposite charges comprising the dipole. In the case of the charge, when one accounts for the clustering virtual charges (from quantum electrodynamics), then the charge is actually a set of composite dipoles, each exhibiting the required broken symmetry. Hence charges and dipoles freely absorb (receive) virtual photon energy from the seething vacuum (particle physics proves this), integrating the absorbed "disintegrated" energy into observable form, and re-emitting real, observable EM energy in all directions, continuously. That this has not yet appeared in the electrical engineering model is strictly the fault of the leaders of the scientific community who have not enforced its inclusion. Consequently, the electrical engineers do not even understand what powers

an EM circuit or system. It is energy from the vacuum, extracted and integrated by the source charge and the source dipole.

22. E.g., see D. K. Sen, Fields and/or Particles, Academic Press, London and New York, 1968, p. viii.
23. T. E. Bearden, "Giant Negentropy from the Common Dipole," Proceedings of Congress 2000, St. Petersburg, Russia, Vol. 1, July 2000 , p. 86-98. Also published in Journal of New Energy, 5(1), Summer 2000, p. 11-23. Also carried on www.cheniere.org and the DoE restricted website http://www.ott.doe.gov/electromagnetic/.
24. F. Mandl and G. Shaw, Quantum Field Theory, Wiley, 1984, Chapter 5.
25. With the broken symmetry of unlike charges and with the Nobel Prize awarded to Lee and Yang in 1957.
26. With the fact that virtual charges of opposite charges cluster around an "isolated" observable charge. One differential piece of the observable charge and one clustering virtual charge of opposite sign comprise a composite dipole. The "isolated source charge" may thus be treated as a set of composite dipoles. The broken symmetry of opposite charges solves the problem for any dipole, and thus for the "isolated" charge as a set of composite dipoles.
27. E. T. Whittaker, "On the Partial Differential Equations of Mathematical Physics," Mathematische Annalen, Vol. 57, 1903, p. 333-355.
28. By use of the conventional closed current loop circuit, where all the current in the external circuit (through the losses and the loads) is returned by a "ground return line" back through the source dipole that is freely extracting the EM energy from the vacuum and radiating it. This "spent" current must be forcibly rammed back up through the source dipole itself, scattering the charges and destroying the dipole. We must then input additional energy to the shaft of the generator to rotate it and restore that dipole.
29. For a good description of the modern vacuum, see I. J. R. Aitchison, "Nothing's plenty: The vacuum in modern quantum field theory," Contemporary Physics, 26(4), 1985, p. 333-391.
30. E.g., see R. Podolny, Something Called Nothing: Physical Vacuum: What Is It?, Mir Publishers, Moscow, 1986, p. 181. In mass units, the energy density of the virtual particle flux of vacuum is on the order of 10^{80} grams per cubic centimeter.
31. Willis E. Lamb Jr. and Robert C. Retherford, "Fine structure of the hydrogen atom by a microwave method," Physical Review, 72(3), Aug. 1, 1947, p. 241-243. Lamb received the 1955 Nobel Prize in physics jointly with Polykarp Kush for experiments measuring the small displacement later called the "Lamb shift" of 0+ne of the energy levels in atomic hydrogen.

32. E.g., see W. T. Grandy Jr., "The Explicit Nonlinearity of Quantum Electrodynamics." in The Electron: New Theory and Experiment, David Hestenes and Antonio Weingartshofer, Eds., Kluwer Academic Publishers, Boston, 1991, p. 149-164. Quoting, p. 150: *"...the energy density associated with the Lamb shift would produce a Poynting vector about three times the total power output of the sun, and a gravitational field disrupting the entire solar system!"* We comment that the reaction is ongoing in a maelstrom of additional such violent interactions, and so — even though *each* of these fierce interactions might be energetic enough to disrupt the entire solar system — the *net summation* is a very tame little thing confined to shifting that single little electron, while still participating in the fluctuations of the vacuum energy elsewhere. The summation of infinite things to leave a manageable finite thing, is well-known and used in particle physics, where it is often referred to as *renormalization*. Simply put, the vacuum energy also has a process for "renormalization".

33. H. B. G. Casimir, "On the attraction between two perfectly conducting plates," presented at a meeting of the Royal Netherlands Academy of Arts and Sciences on 29 May, 1948. Published in the same year in Proceeding Koninklijke Nederlandse Akademie van Wetenschappen, Amsterdam, Vol. 51(7), 1948, p. 793-796.

34. An accurate modern measurement of the Casimir effect is given by S. K. Lamoreaux, "Demonstration of the Casimir Force in the 0.6 to 6μm range," Physical Review Letters, 78(1), Jan. 6, 1997, p. 5-8.

35. Daniel C. Cole and Harold E. Puthoff, "Extracting Energy and Heat from the Vacuum," Physical Review E, 48(2), Aug. 1993, p. 1562-1565.

36. Myron W. Evans, private correspondence.

37. Nikola Tesla, "The True Wireless," Electrical Experimenter, May 1919. Heaviside was speaking of Hertzian (transverse) EM waves in the vacuum. Indeed, with Whittaker's bidirectional longitudinal EM wave decomposition of the scalar potential, and by quantum field theory's finding that the time-polarized photon and longitudinal photon, when combined, are observed as the instantaneous scalar potential, one can make a very good case that the EM waves in space are longitudinal EM waves anyway. This is particularly interesting since nearly a dozen nations have already secretly weaponized longitudinal EM wave technology.

38. Particularly see M. W. Evans, "O(3) Electrodynamics," in Modern Nonlinear Optics, Second Edition, 3 vols., Edited by M.W. Evans, Wylie, New York, 2001, Part 2, p. 79-267. Evans also has more than 600 other papers in the scientific literature.

39. M. W. Evans, "The Link Between the Sachs and O(3) Theories of Electrodynamics," in M. W. Evans (Ed.), Modern Nonlinear Optics, Second Edition, , 3 vols. Wiley, 2001; vol. 2, p. 469-494; M.W. Evans *et al.*,

"Derivation of O(3) Electrodynamics from the Irreducible Representations of the Einstein Group," Foundations of Physics Letters, 2002 (in press).

40. E.g., M. W. Evans *et al.*, "Explanation of the Motionless Electromagnetic Generator with O(3) Electrodynamics," Foundations of Physics Letters, 14(1), Feb. 2001, p. 87-94; — "Explanation of the Motionless Electromagnetic Generator by Sachs's Theory of Electrodynamics," Foundations of Physics Letters, 14(4), Aug. 2001, p. 387-393.
41. M. W. Evans *et al.*, "Anti-Gravity Effects in the Sachs Theory of Electrodynamics," Foundations of Physics Letters, 2002 (in press).
42. Floyd Sweet and T. E. Bearden, "Utilizing Scalar Electromagnetics to Tap Vacuum Energy," Proceedings of the 26th Intersociety Energy Conversion Engineering Conference (IECEC '91), Boston, Massachusetts, 1991, p. 370-375. Sweet's device produced 500 watts for a 33 microwatt input. A highly successful anti-gravity experiment was also performed, and is reported in the paper. Unfortunately Sweet later died and never fully revealed the activation secret by which barium ferrite magnetic materials could be in strong self-oscillation at 60 Hertz. Weak self-oscillation of such permanent magnetic materials at higher frequency is known, of course. E.g., see V. S. L'vov, Wave Turbulence Under Parametric Excitation: Applications to Magnets, Springer Series in Nonlinear Dynamics, Springer-Verlag, New York, 1994.
43. Max Planck, in G. Holton, Thematic Origins of Scientific Thought, Harvard University Press, Cambridge, MA, 1973.
44. Arthur C. Clarke, in "Space Drive: A Fantasy That Could Become Reality" NSS ... AD ASTRA, Nov/Dec 1994, p. 38.
45. Maciej J. Ogorzalek, Chaos and Complexity in Nonlinear Electronic Circuits, World Scientific, New Jersey, 1997, p. vii.
46. S. Austen Stigant, "Gabriel Kron on Tensor Analysis, A bibliographical record," BEAMA Journal, Aug. 1948 gives an extensive bibliography of Kron's publications.
47. J. W. Lynn and R. A. Russell, "Kron's Wave Automaton," Journal Unk, date unk, p. 131. We have the physical paper, with the identifying markings unfortunately obscured. A more complete citation of this reference would be appreciated.
48. David J. Bohm, "A Suggested Interpretation of the Quantum Theory in Terms of 'Hidden' Variables, I and II." Physical Review, 85(2), Jan. 15, 1952, p. 166-179 (Part I); 180-193 (Part II).
49. Indeed, several nations of the word have already highly weaponized Bohm's quantum potential, and such weapons are presently the most decisive weapons on earth. Discussion of this area is beyond the scope of this paper. If one wishes to guess at the nations, one should observe where Bohm went

when the U.S. effectively booted him out of the U.S. Russia and China also have quantum potential weaponry.

50. William K. George, in Editing the Refereed Scientific Journal, edited by Robert A. Weeks and Donald L. Kinser, IEEE Press, New York, 1994, p. 227-228.
51. M. W. Evans, P. K. Anastasovski, T. E. Bearden et al., "Derivation of the B(3) Field and Concomitant Vacuum Energy Density from the Sachs Theory of Electrodynamics," Foundations of Physics Letters, 14(6), Dec. 2001, p. 589-593; — "Anti-Gravity Effects in the Sachs Theory of Electrodynamics," Foundations of Physics Letters, accepted (in press); — "Operator Derivation of the Gauge Invariant Proca and Lehnert Equations: Elimination of the Lorentz Condition," Foundations of Physics, 30(7), 2000, p. 1123-1130; — "Spontaneous Symmetry Breaking as the Source of the Electromagnetic Field," accepted by Foundations of Physics Letters (in press); — "The Effect of Vacuum Energy on the Atomic Spectra," Foundations of Physics Letters, 13(3), June 2000, p. 289-296; — "Runaway Solutions of the Lehnert Equations: The Possibility of Extracting Energy from the Vacuum," Optik, 111(9), 2000, p. 407-409; — "Classical Electrodynamics without the Lorentz Condition: Extracting Energy from the Vacuum," Physica Scripta, 61(5), May 2000, p. 513-517; — "Derivation of the Lehnert Field Equations from Gauge Theory in Vacuum: Space Charge and Current," Foundations of Physics Letters, 13(2), Apr. 2000, p. 179-184.
52. Rolf Landauer, "Fashions in Science and Technology," Physics Today, 50(12), Dec. 1997, p. 62.
53. Derek John de Solla Price, Little Science, Big Science, Columbia University Press, New York, 1963.
54. Secretary of Defense William Cohen at an April 1997 counterterrorism conference sponsored by former Senator Sam Nunn. Quoted from DoD News Briefing, Secretary of Defense William S. Cohen, Q&A at the Conference on Terrorism, Weapons of Mass Destruction, and U.S. Strategy, University of Georgia, Athens, Apr. 28, 1997.
55. Teruo Kawai, "Motive Power Generating Device," U.S. Patent No. 5,436,518. Jul. 25, 1995.
56. The Yakuza also have longitudinal EM wave weapons, and have had them for some time.
57. Tentatively titled Energy from the Vacuum: Concepts and Principles, World Scientific, 2002 (in preparation).

Chapter 20

Market Information

Sure enough, Tesla was able to create the Roadster for just $140 million, compared to the billions of dollars it costs some car companies to design a standard gasoline model.

MACx is not only greener than Tesla, but far more economical to build.

Telsa plans on selling 10,000 cars in 2009 at $60,000, generating $600 million in sales.

No need to worry. Tesla's Model S is due to start production in 2010. The car, which will sport a similar fully electric, 200-plus-mile-range battery, is expected to cost around $60,000. While we're still short on pictures and further details about the Model S, a Tesla representative suggested it'll be roughly the size of a BMW 5 Series -- a roomy, 4-door sedan. Further down the road, Tesla plans on rolling out cars for less than $30,000.

Besides, even if the prices still seem a bit steep, the amount you'll save on gas is staggering. A Tesla brochure I picked up claims the car uses roughly 36 kilowatt-hours of electricity for every 100 miles driven. Energy prices vary throughout the nation, but using an average $0.102 per kilowatt hour, Tesla's efficiency compared to a car getting 20 MPG is equivalent to paying about $0.73 for a gallon of gas, not to mention it's completely emissions-free.

This might be the start of something big

Why is Tesla such a groundbreaking company? It was not only able to create an extraordinarily efficient car, but also did so without compromising any of the customary car elements we've become so attached to. If the next generation of vehicles were slower than a Flintstone car, had to be constantly recharged, and looked like something straight out of Star Trek, people would be hesitant to make the plunge. Tesla's car might rival almost any other in looks and performance, even before you see what's under the hood. That will allow people to hold on to their drive-happy lives, while sticking a knife in the gas crisis and helping the environment at the same time.

Are you as amped about Tesla as I am? Stay tuned. Rumor has it we may see a Tesla IPO sometime in 2009.

U.S. passenger vehicle sales posted another double-digit drop last month, as the shift away from gas-guzzling trucks intensified in May alongside soaring gasoline prices. Sales of light trucks plunged 27% y/y at the 'traditional' Big Three, reducing their market share to a record low of 45.3% and pulling industry volumes down to an annualized 14.3 million units-well below the 15.0 million unit average of the previous four months and the weakest level since the height of the Asian currency crisis a decade ago. Anecdotal evidence suggests a further sharp sales deterioration in June.

With consumers shifting from gas-guzzling vehicles, the Ford F-series, the best-selling vehicle in the United States since 1991, was dethroned by the Honda Civic. Highlighting the shift towards cars and fuel-efficient models, the Toyota Corolla and Camry, as well as the Honda Accord, also outsold

the F-series last month. This prompted Ford to slash its third-quarter North American output 18% below a year earlier and to offer 'employee pricing' on all purchases of the F-series. General Motors has also announced that it will close four assembly plants, including the truck plant in Oshawa,

Canada, and slash second-half truck output by 170,000 units.

In contrast, vehicle purchases remained strong in Canada last month, with volumes nearly matching the monthly record for the month of May set last year. We estimate that sales totaled an annualized 1.70 million units, up from a full-year 2007 total of 1.65 million units, but down slightly from a January-April average of 1.76 million units. Imported brands continued to lead the way, setting an all-time sales record last month, as several automakers, including Toyota, Nissan and Mercedes-Benz, posted double-digit year-over-year gains.

In contrast, volumes at the 'traditional' Big Three fell 9% below a year earlier, reducing their market share to an all-time low of 47%. Light trucks accounted for the weakness, falling 14% y/y alongside a 23% slump in pickup truck volumes. Sales of pickup trucks were particularly weak at General Motors, plunging 37% y/y. Excluding gas-guzzling SUVs and pickup trucks, the Big Three's volumes held up much better, posting a modest 4.5% y/y decline.

DÉJÀ VU – Americans Shift To Fuel-Efficient Vehicles

This year's surge in gasoline prices to a record US$4.00 per gallon has set off a dramatic shift in vehicle buying patterns in the United States, matching the adjustment that occurred in the oil shock of the early 70's. With Americans abandoning their gas-guzzling SUVs and pickup trucks for smaller, more fuel-efficient vehicles, we estimate that the average fuel-efficiency of this year's fleet has climbed by nearly 20% from the previous model year – a level matching the improvement that occurred in 1980, when inflation-adjusted oil prices were approaching $100 per barrel.

Based on sales through May, so far this year, the fuel efficiency of vehicles bought in the United States has increased to an average of 24.4 miles per gallon, up from 20.2 miles in 2007 and an average of 19.8 during the past decade.

With Americans increasingly buying small cars, this segment became the largest in the United States last month, surpassing both mid-size cars and pickup trucks.

Small cars now account for one-quarter of overall U.S. sales (cars & light trucks), up from 16% in 2007 and an average of 14% in the 2002-2006 period. In fact, small cars and fuel-efficient crossover utility vehicles (CUVs) now account for 42% of the U.S. market, up from 30% in 2006, and have doubled their share from as recently as 2001. In contrast, gas-guzzling pickups and SUVs have slumped to only 19% of overall U.S. volumes, down from a peak of 36% in 2001.

Despite the dramatic change in U.S. vehicle buying patterns, the fuel efficiency of the current U.S. vehicle fleet is still more than 40% short of the requirement set out in last year's landmark U.S. energy bill. Under the energy bill, corporate average fuel economy (CAFÉ) requirements are set to climb to a fleet average fuel-economy target of 35 mpg by 2020, with substantial improvements by 2015. Meeting these targets requires not only a further significant shift towards smaller vehicles, but also sizeable new investments by automakers in everything from hybrids to plug-in gas-electric vehicles.

Attracting the investment needed to produce vehicles that meet the new CAFÉ standard may pose a challenge for Canada, because most of our capacity is currently geared towards large vehicles. Canada has the most productive assembly plants in North America, placing four facilities – Oshawa's #1 and #2 car plants, the CAMI operation and Chrysler's car plant in Brampton – among the top ten according to the latest Harbour Report. However, 35% of Canada's total assembly capacity is currently dedicated to producing large cars, minivans and pickup trucks – segments that have posted sales slumps of at least 20% in the United States so far this year. In fact, only the Japanese-owned facilities in Cambridge and Alliston, Ontario are currently producing small fuel-efficient cars in Canada that meet the new 20/20 CAFÉ targets. These facilities account for only 20% of Canada's overall assembly capacity. The challenge is heightened because Canada has become a high-cost jurisdiction, in response to the strong loonie and union contracts, making it more difficult to attract the billions of dollars needed to retool its facilities to meet the new 20/20 CAFÉ standards.

Aside from the shift to smaller cars & light trucks, Americans are also driving less and taking more public transit. Estimates from the U.S. Energy Department indicate that gasoline demand was 4% below a year earlier in mid-June, one of the sharpest fall-offs since the oil shock of the early 70's. In addition, the American Public Transportation Association recently indicated that mass transit use increased 3.3% y/y in the first quarter, led by a double-digit surge in light rail. This reflects the slashing of U.S. household disposable income by more than US$90 billion through the first quarter due to record gasoline prices. We estimate that with oil prices around US$130 per barrel, U.S. household purchasing power will be reduced by an additional $80 billion in the second quarter, nearly 1% of disposable income, a level approaching the reductions that occurred in the oil shock of the 70's.

Faced with gasoline prices of $1.30 per liter, Canadians are also buying more fuel-efficient vehicles. However, the shift is more modest domestically, given that the Canadian market has always been dominated by small cars. The Canadian vehicle fleet is, on average, about 50% more fuel-efficient than the U.S. fleet. This has limited this year's improvement in the fuel-efficiency north of the 49th parallel to a modest 3%.

International Car Sales Outlook

	1990-99	Peak 2000	2001-04	2005-06	2007	2008f
			(millions of units)			
TOTAL SALES	39.20	46.64	47.02	51.75	54.76	56.29
North America*	16.36	19.77	19.44	19.45	18.83	17.49
Canada	1.27	1.55	1.60	1.60	1.65	1.69
United States	14.55	17.35	16.85	16.72	16.09	14.70
Mexico	0.54	0.87	0.99	1.13	1.09	1.10
Western Europe	13.11	14.75	14.49	14.65	14.75	14.60
Germany	3.57	3.38	3.27	3.41	3.15	3.18
Eastern Europe	1.18	2.38	2.13	2.80	3.58	4.01
Russia	0.78	1.03	1.09	1.42	2.31	2.66
Asia	6.91	7.85	9.14	12.33	14.29	15.22
China	0.33	0.61	1.55	3.68	5.15	5.92
India	0.31	0.60	0.66	0.94	1.18	1.30
South America	1.64	1.89	1.82	2.52	3.31	3.97
Brazil	0.94	1.17	1.22	1.46	1.98	2.37

*Includes light trucks.

Chapter 21

Competition

Presently, there are a number of companies working towards developing a production vehicle. The main problem is funding and providing a product that the public will purchase.

- Tesla Motors (sports cars)
- Wrightspeed (sports cars and plug-in drivetrains for trucks)
- Fisker Automotive (electric sedans)
- Zap (low-speed and sports cars)
- Miles Automotive (low speed)
- Zenn Motors (low speed)
- AC Propulsion (retrofitting Scions for electric)
- Phoenix Motorcars (SUVs)
- Aptera (three-wheelers)
- Porteon (low speed electrics)
- Lightning (sports cars in England)
- Reva (economy cars)
- Ultramotor (electric trishaws)
- Myers Motors (freakish three-wheelers featured in Goldmember)
- Think (electric economy cars)
- Venture Vehicles (three-wheeled electric cars)
- NASA (air bearing turbocharger – heat – gasoline. NASA believes that the air bearing turbochargers are creating a turbomachinery revolution. NASA believes that turbocharger companies in the future will be using air bearings, with the low-temperature coatings, for possible use in automobiles, trucks, boats and aircraft engines.)
- Toyota (air bearing turbocharger – heat – gasoline. A turbocharger

with the improved bearing has been run at a rotational speed up to 200,000 rpm.)

- Air Car in France (pushing pistons - refill every 160km)
- EEstor in Texas (capacitor w/gasoline)
- Car in UK (electric motor w/gasoline)

Chapter 22

Forces of Nature

This section is mostly about tornadoes, but I'm also throwing in the hurricanes because of the similarity of form and to show the action of the coriolis force which plays some part in the formation of tornadoes. Most, though ***not all,*** tornadoes rotate with the coriolis expected handedness.

Hurricanes are quite well understood in their basic anatomy. The basic steps of formation:

1. Ocean temperature rises
2. Airs warms
3. Water evaporates
4. Warm air rises
5. Cooler air rushes in to take it's place
6. Coriolis force causes rotation.
7. Energy source removed (hot water)
   Storm dies [as over land]

That's not what a tornado is. In this case, a warm air mass goes up and over a cold air mass (it's generally lighter). Friction between the two creates a roll of rotating air and this tips over and becomes a column which may touch down (as when it ***sees*** a trailer park).

The above explains nothing in functional terms. So ...

What is an air mass?

One place is hot, another cold (or dry or humid or any combination). There is a boundary between them. One might then inquire as to why the two don't mix. After all, it's just air and air mixes easily from anyone's experience, doesn't it?

Yes and no. Our experience is with room-sized bodies of air, and these do mix quite well. However, it takes ***some*** amount of time. If the room were the size of Connecticut, the time needed to mix completely is enormously longer than a room. It's the same scenario at basically the same temperatures, so it mixes at a fixed rate (fixed in the sense of the 'big picture').

Therefore, for large areas, the dividing line between two differing areas might as well be solid given the small percentage of mixing that occurs over, say, two or three days time. If you've ever been in a jet, you may have seen "cliffs" of clouds showing obvious demarcation between "*air masses*".

What happens when two different masses collide?

There is friction between the cold and hot masses and the roll develops. This constitutes another "air mass" (differentiated by its rotation aspect). Note here also that this new mass is lighter (less dense = generally lower air pressure) than the cold air mass.

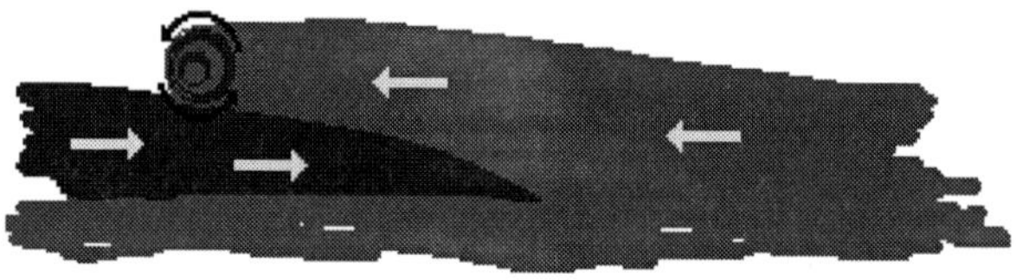

Roll mechanics

Friction at the hot/cold boundary slows relative air velocity at that point, creating a very low- pressure area in front of where the roll will form. This causes the "over the top air" to rush ahead even faster to fill that "vacuum". The air on the top of the roll is therefore faster than on the bottom and the whole thing is pushed along by the over the top air mass at no less than it's velocity. As a result, the roll has a rotation velocity faster than the relative velocity of the colliding air masses.

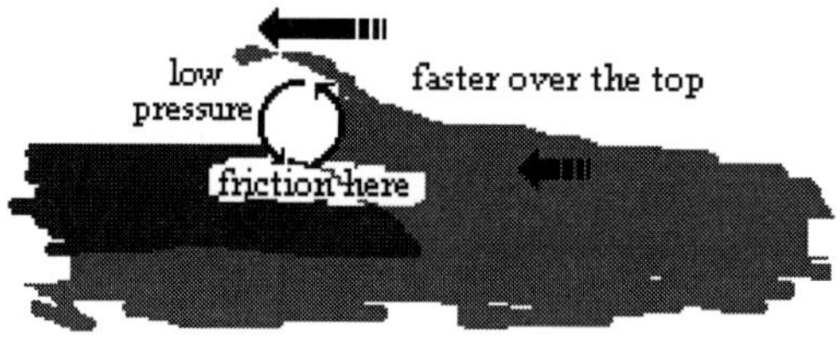

By itself, this "roll" is harmless. We need a mechanism to get it to the ground. It needs to be tipped over 90 degrees as a single unit and brought down to the trailer park ground.

The mechanism involved is ***not*** electromagnetism.

What mechanism then?

All the lighting and thunder is mere storm dressing ... all sound and fury signifying nothing (like the head gear worn by a Las Vegas showgirl). We're just interested in the "body" of the mechanism. The true mechanism is simple air pressure consolidations and a third collision.

We need another air mass collision parallel to the ground but perpendicular to the "roll". Something to give it a "turning moment". then ... it will precess like a top and get pointed in the right direction - at the ground trailer park.

Now, you might think, "Precess like a top? Why doesn't it just "disintegrate"?

Because to disintegrate, it would have to change shape. For instance, what if it just precessed in small slices like a "slinky"?

No good ...

The volume of the cylinder laid down in this manner is less than the former. The gas must then be compressed, but there is no force to do compression at this stage. Therefore, it may only rotate as an integrated, independent *"air mass"*.

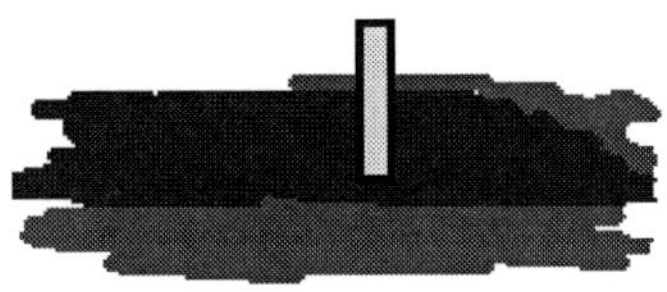

Once rotated, even though it has less weight than a comparably sized air mass near the ground, it will still "float" ***on and in*** that mass like a heavier-than-water bar with a cork on one end. Some of it will extend down to ground level and bob up and down wherever "atmospheric waves" may take it.

As it drops down, it should have a funnel shape because the lower, denser air mass is compressing the "vortex". When it stops dropping, the funnel shape should disappear into a smooth walled cylinder at maximum compression.

Note: Angular momentum equals **mvr**, so since the **m**ass is constant, the **v**elocity of rotation must increase in proportion to the decrease in **r**adius. If the funnel is compressed to 1/4 its original diameter, it will have a velocity 4 times the original [if 60 mph at its initial "roll", then 240 mph after compression].

This model requires three things for a tornado to occur

The hot air, the cold air mass and another air mass (*of undetermined size & type ... possibly just an asymmetry in the cold or hot air mass causing the roll to stick a little here and there*) skewing the rolled air to make it "stand up" like a top.

As air in the storm roils about, there will be swirls where the t-roll will sit comfortably, i.e., rotate in the same direction. Most swirls will tend to be affected by coriolis forces (as in hurricanes and water going down your tub drain) so that non-reinforced rotations will be less likely to occur, hence, the (non-exclusive) preference for counter-clockwise tornadoes in the northern hemisphere.

Why don't tornadoes disintegrate by friction with the surrounding air?

They do. They don't last terribly long. Not like a hurricane. However, it does take some time ... after all it's a gas-gas interface, i.e., pretty friction free. They can be stopped much sooner, however ... if you put lots of aluminum trailers in their path.

Chapter 23

Marketing and Advertisements

1. Movie shows a hurricane force picking up everything in its path. A Magnetic Air Car™ drives-by and the hurricane jumps into the car as the car continues on its way effortlessly with the power of the hurricane operating under its hood.
2. Smog is shown in an out-of-control city horizon. A Magnetic Air Car™ drives-by and the scenery changes to that of a Microsoft Blue Clouds Logo. Magnetic Air Car™ technology creates sky clean cold air and breathable atmosphere for a truly green tech solution.
3. People are storming the White House violently complaining about the high cost of fuel and the economy. When a Magnetic Air Car™ drives-by, the crowd completely calms down and disperses immediately to surround the car in awe. Even the cops come over.
4. Magnetic Air Car™ drives-by with the whistling sound of cool air breezing. Like the sounds of the low vibrations of compressors in a tire shop as the car accelerates.
5. Mel Gibson in the movie Braveheart is crying out "FREEDOM" with his last dying breath as the Magnetic Air Car™ goes by.
6. Charleston Heston is talking in the character voice of Moses. He quotes Isaiah chapter 2 verse 4: "They shall beat their swords into plowshares, And their spears into pruning hooks; Nation shall not lift up sword against nation, Neither shall they learn war anymore." as the Magnetic Air Car™ goes by.
7. Christopher Cross's song, Ride Like TheWind is playing in the background. A shot of the Magnetic Air Car™ is shown racing in the dessert salt flats in Nevada.
8. Headline News, PBS correspondent Glenn Beck stated that air cars are WORTH BILLIONS and that investors are coming in by the minute.

9. In the movie I, Robot; starring Will Smith. The co-star is riding on the back of an ancient motorcycle. She cries out, "This thing doesn't run on gasoline. Gasoline is dangerous."
10. Senate Hearing - Steve being cross examined by Senator Kennedy asking why Steve is wasting investors and taxpayer's money flying in Wash D.C. Don't you have a fiduciary responsibility to use your funds more conservatively? yes Sir. I flew into Wash DC in our prtototype airplane. We didn't use any fuel. We cleaned the air while the plane was in use. I think that's being responsible... don't you think?
11. In the movie, "The Day the Earth Stood Still", Keenu Reeves, the alien from another galaxy tells the earth woman that mankind is about to be destroyed because man is polluting the earth.

Keenu Reeves replies, "If you live, the earth dies, if you die the earth lives."

The earth woman replies, "We can change, we can change!" She sounded desperate.

We can then Cut to a Magnetic Air Car driving by cleaning up the atmosphere!!!!!!!!

At the end of the film Keenu Reeves replies, "Maybe there is hope for you yet!"

This may be a perfect Commercial for the Magnetic Air Car in our future!

CHAPTER 24

MAGNETIC MINING

There are deposits in northern China in the conquered Mongolian territory called Inner Mongolia. That is the location of most of the current mining of rare earth metals. These were mostly created hydro thermally in rising water columns over a long time. These also are relatively hard to find.

A much larger area containing a huge volume is located in huge lateritic (clay) deposits in southern China. These are barely being used currently. We can set up operations where a circular excavation bucket wheel operates in a field, putting the material onto conveyor belts to be transported to a processing plant. There the metals are extracted and concentrated. Specific hydrocarbons are used to attract specific elements and then the hydrocarbons are boiled off and then distilled and reused with over 99% efficiency. The second major source of Chinese REE is ion-adsorption ores in lateritic weathering crusts developed on granitic and syenitic rocks in tropical southern China.

Required elements are ND, Pr, Dy, Sm and Gd and we are interested in iallal of the rare earth metals. They (Nd, Pr, Dy, Sm and Gd) comprise a little under half of the elements that are available in that ore. We suggest to take the rest of the ore after concentrating it into their oxides. The 5 elements Nd, Pr, Dy, Sm and Gd should actually be converted into their metal form.

The bucket wheel excavators can be made in Ukraine for a fraction of the cost that Koreans and Japanese companies sell their equipment. One Chinese company copied the competition and now makes competing products. They tend to break often but may be cheaper than the Ukrainian at least initially. They may be now importing Ukrainian, German or Finish steels that have the required qualities to make a superior machine. Their web site is http://jsneoplant.en.ecplaza.net. The company is JS NEOPLANT CO., LTD and a contact there is Mr. Charles, Jim at 86-21-5422-3041 and with an address of No.64-102,633Nong ,Wuzhong R/DShanghaiai, Shanghai, China 201103.

Elgen motors, designed and manufactured by Terry Sosansk out of Texas run at 600V (volts) and about 130A (amps) for a total power of 78KW. Elgen generators supplied this power in 2000 and the tests went on for 7 years. The Ukrainian excavators typically run on 380V. The Elgen generators have run Ukrainian excavators without many problems. They call it a "nuclear reactor" and keep people away by putting radiation symbols on the inside packing of the machine.

The bucket wheel excavators with capacity of 3600 m3 per hour can supply adequate raw material for about 750,000 tons per year of rare earth metals. We would need about 12,000,000 tons per year for 20 years to replace 1/2 of the power in the USA to magnetic power. That is 15 excavators working 24 hours a day for 20 years. Obviously for cars, this number is about doubled. Refrigeration is another large area for these rare earth metals. For the world the number is even greater. The current rare earth metal production is over 100,000 tons a year. We need to ultimately increase it 1000 times.

Rare Earth Elements --Critical Resources for High Technology

High-technology and environmental applications of the rare earth elements (REE) have grown dramatically in diversity and importance over the past four decades. As many of these applications are highly specific, in that substitutes for the REE are inferior or unknown, the REE have acquired a level of technological significance much greater than expected from their relative obscurity. Although actually more abundant than many familiar industrial metals, the REE have much less tendency to become concentrated in exploitable ore deposits. Consequently, most of the world's supply comes from only a few sources. The United States once was largely self-sufficient in REE, but in the past decade has become dependent upon imports from China.

The rare earth elements (REE) form the largest chemically coherent group in the periodic table. Though generally unfamiliar, the REE are essential for many hundreds of applications. The versatility and specificity of the REE has given them a level of technological, environmental, and economic importance considerably greater than might be expected from their relative obscurity. The

United States once was largely self-sufficient in these critical materials, but over the past decade has become dependent upon imports (fig. 1). In 1999 and 2000, more than 90% of REE required by U.S. industry came from deposits in China.

Global rare earth element production (1 kt=10^6 kg) from 1950 through 2000, in four categories: United States, almost entirely from Mountain Pass, California; China, from several deposits; all other countries combined, largely from monazite-bearing placers; and global total. Four periods of production are evident: the monazite-placer era, starting in the late 1800s and ending abruptly in 1964; the Mountain Pass era, starting in 1965 and ending about 1984; a transitional period from about 1984 to 1991; and the Chinese era, beginning about 1991.

Although the 15 naturally occurring REE are generally similar in their geochemical properties, their individual abundances in the Earth are by no means equal. In the continental crust and its REE ore deposits, concentrations of the most and least abundant REE typically differ by two to five orders of magnitude. As technological applications of REE have multiplied over the past several decades, demand for several of the less abundant (and formerly quite obscure) REE has increased dramatically.

The diverse nuclear, metallurgical, chemical, catalytic, electrical, magnetic, and optical properties of the REE have led to an ever increasing variety of applications. These uses range from mundane (lighter flints, glass polishing) to high-tech (phosphors, lasers, magnets, batteries, magnetic refrigeration) to futuristic (high-temperature superconductivity, safe storage and transport of hydrogen for a post-hydrocarbon economy).

Some Applications of the Rare Earth Elements.

Many applications of REE are characterized by high specificity and high unit value. For example, color cathode-ray tubes and liquid-crystal displays used in computer monitors and televisions employ europium as the red phosphor; no substitute is known. Owing to relatively low abundance and high demand, Eu is quite valuable—$250 to $1,700/kg (for Eu_2O_3) over the past decade.

Fiber-optic telecommunication cables provide much greater bandwidth than the copper wires and cables they have largely replaced. Fiber-optic cables can transmit signals over long distances because they incorporate periodically spaced lengths of erbium-doped fiber that function as laser amplifiers. Er is used in these laser repeaters, despite its high cost (~$700/kg), because it alone possesses the required optical properties.

Specificity is not limited to the more exotic REE, such as Eu or Er. Cerium, the most abundant and least expensive REE, has dozens of applications, some highly specific. For example, Ce oxide is uniquely suited as a polishing agent for glass. The polishing action of CeO_2 depends on both its physical and chemical properties, including the two accessible oxidation states of cerium, $Ce,^{3+}$ and Ce^{4+}, in aqueous solution. Virtually all polished glass products, from ordinary mirrors and eyeglasses to precision lenses, are finished with CeO_2.

Table 1. Names and symbols of the REE

La	lanthanum	Tb	terbium
Ce	cerium	Dy	dysprosium
Pr	praseodymium	Ho	holmium
Nd	neodymium	Er	erbium
Pm	promethium	Tm	thulium
Sm	samarium	Yb	ytterbium
Eu	europium	Lu	lutetium
Gd	gadolinium	Y	yttrium

Permanent magnet technology has been revolutionized by alloys containing Nd, Sm, Gd, Dy, or Pr. Small, lightweight, high-strength REE magnets have allowed miniaturization of numerous electrical and electronic components used in appliances, audio and video equipment, computers, automobiles, communications systems, and military gear. Many recent technological innovations already taken for granted (for example, miniaturized multi-gigabyte portable disk drives and DVD drives) would not be possible without REE magnets.

Environmental applications of REE have increased markedly over the past three decades. This trend will undoubtedly continue, given growing concerns about global warming and energy efficiency. Several REE are essential constituents of both petroleum fluid cracking catalysts and automotive pollution-control catalytic converters. Use of REE magnets reduces the weight of automobiles. Widespread adoption of new energy-efficient fluorescent lamps (using Y, La, Ce, Eu, Gd, and Tb) for institutional lighting could potentially achieve reductions in U.S. carbon dioxide emissions equivalent to removing one-third of the automobiles currently on the road. Large-scale application of magnetic-refrigeration technology (described below) also could significantly reduce energy consumption and CO_2 emissions.

Chemical periodic table delineating the 16 rare earth elements (REE): the lanthanides, La through Lu, plus Y, whose geochemical behavior is virtually identical to that of the heavier lanthanides. Promethium has no long-lived isotopes and occurs naturally on Earth only in vanishingly small quantities. An represents the first 14 actinide elements; Lr is the last actinide.

In many applications, REE are advantageous because of their relatively low toxicity. For example, the most common types of rechargeable batteries contain either cadmium (Cd) or lead. Rechargeable lanthanum-nickel-hydride (La-Ni-H) batteries are gradually replacing Ni-Cd batteries in computer and communications applications and could eventually replace lead-acid batteries in automobiles. Although more expensive, La-Ni-H batteries offer greater energy density, better charge-discharge characteristics, and fewer environmental problems upon disposal or recycling. As another example, red and red-orange pigments made with La or Ce are superseding traditional commercial pigments containing Cd or other toxic heavy metals. The next high-technology application of the REE to achieve maturity may be magnetic refrigeration. The six REE ions GD3+ through Tm3+ have unusually large magnetic moments, owing to their several unpaired electrons. A newly developed alloy, GD(5) (Si(2) Ge(2), with a "giant magne-tocaloric effect" near room temperature reportedly will allow magnetic refrigeration to become competitive with conventional gas-compression refrigeration. This new technology could be employed in refrigerators, freezers, and residential, commercial, and automotive air conditioners. Magnetic refrigeration is considerably more efficient than gas-compression refrigeration and does not require refrigerants that are flammable or toxic, deplete the Earth's ozone layer, or contribute to global warming.

Element Abundances

"Rare" earth elements is a historical misnomer; persistence of the term reflects unfamiliarity rather than true rarity. The more abundant REE are each similar in crustal concentration to commonplace industrial metals such as chromium, nickel, copper, zinc, molybdenum, tin, tungsten, or lead. Even the two least abundant REE (Tm, Lu) are nearly 200 times more common than gold. However, in contrast to ordinary base and precious metals, REE have very little tendency to become concentrated in exploitable ore deposits. Consequently, most of the world's supply of REE comes from only a handful of sources.

Differences in abundances of individual REE in the upper continental crust of the Earth represent the super-position of two effects, one nuclear and one geochemical. First, REE with even atomic numbers ($_{58}$Ce, $_{60}$Nd, ...) have greater cosmic and terrestrial abundances than adjacent REE with odd atomic numbers ($_{57}$La, $_{59}$Pr, ...). Second, the lighter REE are more incompatible (because they have larger ionic radii) and therefore more strongly concentrated in the continental crust than the heavier REE. In most rare earth deposits, the first four REE—La, Ce, Pr, and Nd—constitute 80 to 99% of the total. Therefore, deposits containing relatively high grades of the scarcer and more valuable heavy REE (HREE: Gd to Lu, Y) and Eu are particularly desirable.

Prices and abundances of the rare earth elements. Prices are for 1999 or 2000 in U.S. dollars per kilogram of REE metal, in two forms: (1) as oxides, in 2 to 25 kg packages, at 95 to 99.99% purity; (2) as 0.1 to 0.45 kg metal ingot, at 99.9% purity. Two representative REE ores—high-grade carbonatite ore from Mountain Pass, California, and lateritic ion-adsorption ore from southern China—are compared with Earth's upper continental crust. Z, atomic number.

Abundance (atom fraction) of the chemical elements in Earth's upper continental crust as a function of atomic number. Many of the elements are classified into (partially overlapping) categories: (1) rock-forming elements (major elements in green field and minor elements in light green field); (2) rare earth elements (lanthanides, La–Lu, and Y; labeled in blue); (3) major industrial metals (global production >~3x10^7 kg/year; labeled in bold); (4) precious metals (italic); and (5) the nine rarest "metals"—the six platinum group elements plus Au, Re, and Te (a metalloid).

Rare Earth Element Resources

From the discovery of the REE (during the period 1794–1907) through the mid-1950s, a few of the REE were produced in modest amounts from monazite-bearing placers and veins, from pegmatites and carbonatites, and as minor byproducts of uranium and niobium extraction. During this time, the middle and heavy REE generally were available in pure form only in sub-kilogram quantities and were chiefly chemical curiosities.

In 1949, a carbonatite intrusion with extraordinary contents of light REE (8 to 12% rare earth oxides [REO]), was discovered at Mountain Pass, in the upper Mojave Desert, California (fig. 5).

The REE at Mountain Pass are hosted chiefly by bastnäsite, (Ce,La,Nd,...) CO_3F, and related minerals. By 1966, this single, world-class deposit (owned and operated by Molycorp, Inc.) had become the paramount source of REE. Early development was supported largely by the sudden demand for Eu created by the commercialization of color television. Mountain Pass, with an average grade of 9.3% and reserves of 20 million metric tons (Mt) REO (at 5% cutoff), remains the only large ore deposit mined solely for its REE content. Mountain Pass ore is very strongly dominated by the light REE (figs. 3, 6). Nonetheless, the large quantities of ore processed and development of solvent-extraction techniques for large-scale separation of individual REE from one another allowed recovery of several of the middle REE. Increased availability led in turn to applications for these formerly exotic elements.

From 1965 through the mid-1980s, Mountain Pass was the dominant source of REE, and the United States was largely self-sufficient in REE. Since 1985, production of REE in China has increased dramatically. Chinese REE production comes chiefly from two sources. The most important is the Bayan Obo iron-niobium-REE deposit, Inner Mongolia. This deposit has geological affinities both to carbonatite REE deposits and to hydrothermal iron-oxide (-Cu-Au-REE) deposits, such as Olympic Dam, Australia, and Kiruna, Sweden. Grades at Bayan Obo are 3 to 6% REO; reserves are at least 40Mt, possibly considerably more. The second major source of Chinese REE is ion-adsorption ores in lateritic weathering crusts developed on granitic and syenitic rocks in tropical southern China. These oxide ores are advantageous in their relatively high proportions of HREE and, especially, in the ease with wihic they can be mined and the REE extracted.

The number of workable REE deposits, already severely limited by the geochemical properties of the REE, has in recent years also been affected by environmental and regulatory factors. Monazite, the single most common REE mineral, generally contains elevated levels of thorium. Although Th itself is only weakly radioactive, it is accompanied by highly radioactive intermediate daughter products, particularly radium, that can accumulate during processing. Concern about radioactivity hazards has now largely eliminated monazite as a significant source of REE and focused attention on those few deposits where the REE occur in other, low-Th minerals, particularly bastnasite.

Rare Earth Element Supply Issues

Over the past several years the only domestic source of REE, the mine at Mountain Pass, California, has operated below capacity and only intermittently. Following environmental and regulatory problems with the main wastewater pipeline, the REE separation (solvent extraction) plant was shut down. Mountain Pass currently produces only bastnäsite concentrates and sells separated REE only from stockpiles produced before the shutdown. Even after the regulatory situation has been resolved, however, the long-term viability of Mountain Pass as a supplier of separated REE for high-technology applications is threatened by market factors.

The Mountain Pass rare earth element mine, Mojave Desert, California, with Clark Mountain (part of the Mojave National Preserve) in the background and Interstate 15 just beyond the hills in the foreground.

In 1999 and 2000, nearly all (more than 90%) of the separated REE used in the United States was imported either directly from China or from countries that imported their plant feed materials from China. The surprisingly rapid progression from self-sufficiency prior to about 1990 to nearly complete dependence on imports from a single country today involves a number of causative factors. These include much lower labor and regulatory costs in China than in the United States; continued expansion of electronics and other manufacturing in Asia; the favorable number, size, and HREE content of Chinese deposits; and the ongoing environmental and regulatory problems at Mountain Pass. China now dominates world REE markets (fig. 1), raising several important issues of REE supply for the United States:

1. The United States is in danger of losing its longstanding leadership in many areas of REE technology. Transfer of expertise in REE processing technology and REE applications from the United States and Europe to Asia has allowed China to develop a major REE industry, eclipsing all other countries in production of both ore and refined products. The Chinese Ministry of Science and Technology recently announced a new national basic research program. Among the first group of 15 high-priority projects to be funded was "Basic research in rare earth materials" (Science, Dec. 18, 1998, p. 2171).
2. United States dependence on imports from China comes at a time when REE have become increasingly important in defense applications, including jet fighter engines and other aircraft components, missile guidance systems, electronic countermeasures, underwater mine detection, antimissile defense, range finding, and space-based satellite power and communication systems.
3. Availability of Chinese REE to U.S. markets depends on continued stability in China's internal politics and economy, and its relations to other countries.
4. Although the present low REE prices caused by abundant supply from China are stressing producers, particularly Mountain Pass, low prices will also stimulate development of new applications. For example, a recent chemistry text notes that "... for many years the main use of lutetium was the study of the behavior of lutetium ..." Several

promising applications for Lu are known, but most are precluded by high cost. If the price of Lu were to decrease from many thousand to a few thousand dollars per kilogram, additional high-technology applications of even this least abundant of the REE undoubtedly would follow. How large a role the United States will play in future expansion of REE technology and markets remains an important, but open, question.

Proportions of individual REE in two representative ores: bastnäsite, dominated by La, Ce, and Nd, with Eu through Lu plus Y totaling only 0.4%; and lateritic ion-adsorption ore, Y-dominated. Dark blue and light blue sectors represent lanthanides of even and odd atomic number, respectively (see figs. 2, 3). Yttrium is indicated by green.

The rare earth elements are essential for a diverse and expanding array of high-technology applications, which constitute an important part of the industrial economy of the United States. Long-term shortage or unavailability of REE would force significant changes in many technological aspects of American society. Domestic REE sources, known and potential, may therefore become an increasingly important issue for scientists and policymakers in both the public and private sectors.

Article contributed by Terry Sosansk

Chapter 25

Patent Application for Magnetic Air Car

12344242
Filed on December 25, 2008

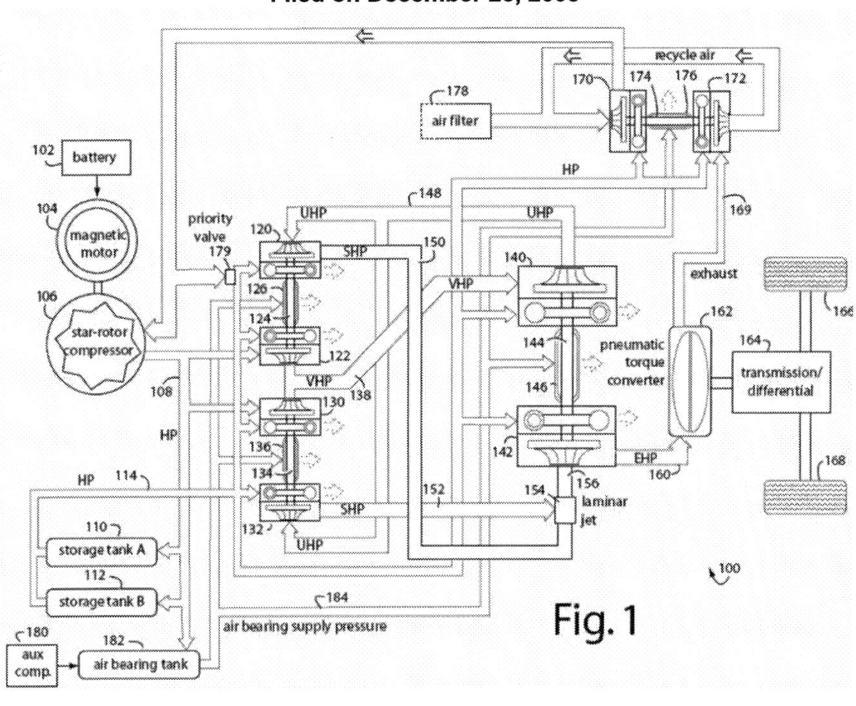

Fig. 1

Abstract of The Disclosure

A magnetic air car uses a magnetic motor to compress input air and save moderately compressed high-pressure air in storage tanks. The compressor and storage tanks deliver the high-pressure working air and operational flows to several stages of compressors that boost the pressures during driving to very high-pressure, then ultra high pressure, then super high-pressure, and finally to extremely high-pressure. A pneumatic torque converter uses jets of the extremely high-pressure to turn an input shaft of a transmission and differential. These, in turn, drive the powered wheels of a car. The compressors float a connecting shaft with matching vanes and impellers on opposite ends on air bearings to reduce shaft turning friction to near zero.

The balance of forces between the two ends of a coupled turbo pair allows a simple air bearing design to operate safely and reliably at high rotational speeds.

Fig. 2

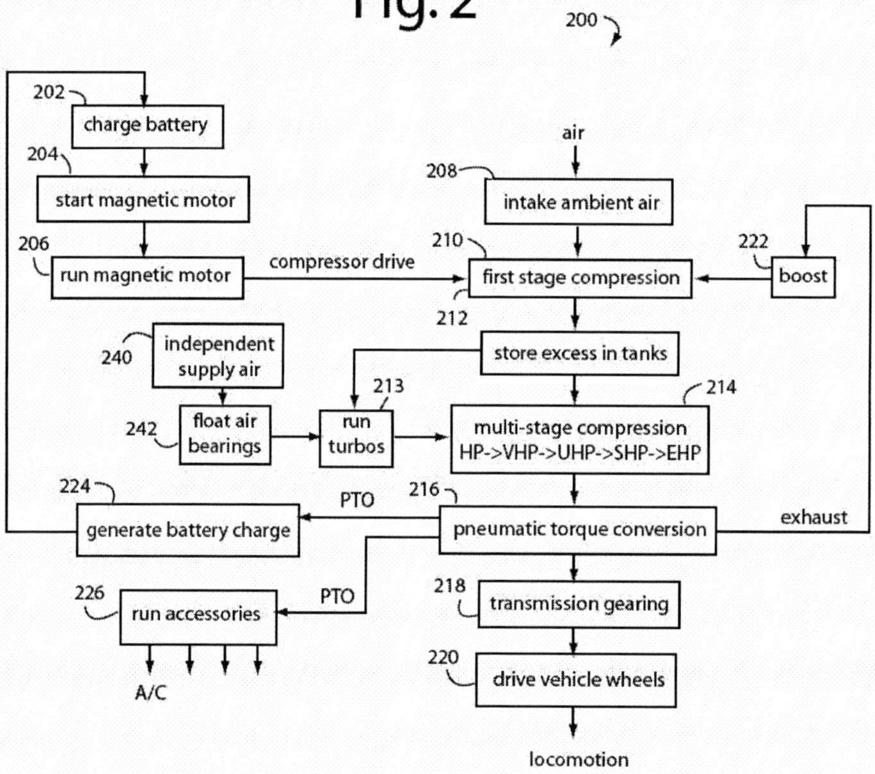

MLF 697-01
MAGNETIC AIR CAR

1. Field of the Invention

The present invention relates to automobiles, and more particularly to propulsion systems for automobiles that compress air on-board with magnetic motors and use the compressed air from storage tanks and a compressor to propel the vehicle.

2. Description of the Prior Art

Gasoline and diesel internal combustion, steam, electric, compressed air, and turbine engines have all been used to propel various kinds of cars and other vehicles.

The range limitation of these vehicles depends on how much fuel or energy can be carried on-board until the next fill-up or recharge. The engine efficiencies also figure into the range, the more efficient the engine the greater will be the range.

New classes of hybrid vehicles are beginning to emerge that have short ranges that can be extended if need be by a small on-board generator. For example, the Chevy Volt plug-in electric car is anticipated to have a range of forty miles on an overnight electric charge of its lithium-ion batteries. Such range is expected to satisfy the needs of over seventy-five percent of America's daily commuters. A small gasoline-powered electric generator on-board automatically starts up if a trip farther than forty miles exhausts the batteries. Liquid fuels like gasoline, kerosene, and diesel, have a relatively high energy content for their weight and volumes, and are therefore the most common kinds of fuels used in cars, trucks, and buses. Internal combustion engines can convert these fuels into mechanical energy, but they are notoriously inefficient and throw away most of the energy as wasted heat. Less than twenty percent of the energy in the fuel actually drives the wheels. Compressed gases have been used too, such as compressed hydrogen or propane, which are burned as fuel. Compressed air can be used directly to run an air pump backwards, where the release of pressure drives a piston that turns the wheels. The mechanics are similar to a steam engine without the boilers. The problem with compressed gases has been the exceedingly high-pressures that have been necessary to

make them useful in vehicles. The usual pressures can exceed four thousand PSI. This makes building the storage tanks to contain these pressures very expensive, and the tanks themselves are very dangerous if exposed to collisions or fire. Damaged super-compressed gas tanks can blevy and rocket like a wild missile. The problem with electric cars is they must carry batteries large enough to power the vehicle. Batteries are inherently heavy, and exotic lightweight batteries are very expensive. The materials used in batteries are often dangerous, e.g., the sulphuric acid in lead-acid cells. The materials are also environmental harmful, e.g., the lead in lead-acid cells. And a charged battery can generate tremendous heat and explode if electrically shorted, like can occur if punctured or warped in a car collision. Permanent magnets express magnetic fields constantly, while electro-magnets express their magnetic fields only so long as current is applied and in proportion to the magnitude of the current flowing through their coils. Electric motors use permanent magnets and electro-magnets in their rotors and stators to turn the rotor and produce a mechanical output from an electrical input. The magnets are switched on-and-off according to their relative positions tokeep the magnetic fields working in the right directions to produce the rotating torque. Magnetic fields between respective north and south poles can be warped and steered by shunting 5
them through highly permeable plates, bars, and shutters. The magnetic fields circulating about a permanent magnet can be effectively switched on-and-off from the perspective of a nearby point by moving or manipulating interposing magnetic shunts. This is one basis for hundreds of different types of permanent magnet electrical generators and reciprocating and rotating magnetic motors. For example, an early magnetic motion conversion device is described by Gerald Howard in United States Patent 3,967,146, issued 06/26/1976. Allan Limb describes a magnetic engine in United States Patent Application 2005/0116567, published 06/02/2005. Michael Cristoforo describes a magnetic force reciprocating motor in United States Patent Application 2008/0122299, published 05/29/2008. A generator with reciprocating and rotating permanent magnets is described by Stephen Kundel in United States Patent 7,400,069, issued 07/15/2008. Harry Sprain describes a magnetic generator in United States Patent 7,265,471, issued 09/04/2007. Using electro-magnetism to drive pistons up and down in a reciprocating motor is described

by Shimon Elmaleh in United States Patent 7,105,958, issued 09/12/2006; and also by Leland Gifford in United States Patent 5,457,349, issued 10/10/1995. Albert Schumann describes a permanent magnet motion conversion device in United States Patent 4,300,067, issued 11/10,1981. Shielding plates are moved into and out of positions in front of a set of stationary magnets. A carriage then shuttles back and forth.

A whole vehicle with a magnetic engine is described by Charles Wortham in United States Patent 5,219,034, issued 06/15/1993. It uses magnetic pistons with electromagnets fitted in the cylinder heads. An AC generator is attached to the rear wheels to recover electrical energy which is then used to charge a battery. Compressed air cars use the expansion of compressed liquid air into gas to drive the pistons in a pneumatic engine. The idea is not new. Prototypes were built in the 1920's and the ideas were discussed in popular mechanics magazines. Modern torpedo weapons and missile launchers use compressed air as well. Conventional compressed air cars typically use high-pressure tanks of 30 MPa (4500 psi or 300 bar). Carbon-fiber tanks can reduce the weight and still have the necessary strength. The Zero Pollution Motors (New Paltz, NY) division of Moteur Development International (France) is presently marketing a ZPM Air Vehicle that promises almost zero pollution motoring. See, http://zeropollutionmotors.us/.

Their vehicle uses a compressed air engine (CAE) with pistons powered by the expansion of electronically injected compressed air. The CAE has an active chamber and two opposing cylinders in each module. A proprietary connection rod keeps the pistons at top dead center during 70° of crankshaft rotation, in order to maintain the required pressure in the cylinder long enough to get maximum work output. The basic engine modules can be coupled together in groups of four or six cylinders for a range of uses. Each compressed air tank in the ZPM vehicle holds 3,200 ft3 of compressed air at 4,500 psi (310 bar). An on-board plug-in electric compressor is used between trips to generate 812 ft3 of compressed air per hour to recharge the tanks, e.g., four hours or overnight.

The first production units are scheduled for deliveries in 2009. At speeds over 35-mph, the 75-hp (56 kW) CAE uses a compressed air multiplier (CAM) to burn small amounts of fuel, e.g., gasoline, propane, ethanol or bio fuels inside a heating chamber to heat the ambient air 5 as it enters the engine. This is said to produce insignificant emissions of only 0.14 pounds of carbon dioxide per vehicle mile. The fuel economy is expected to be 106-miles per gallon. With an eight-gallon fuel tank, the car is calculated have a range of 848 miles. Guy Negre describes a similar engine in United States Patent 7,296,405, issued 11/20/2007, and also United States Patent Application 2007/0101712, published 05/10/2007. The whole car like this is described by him and Cyril Negre in United States Patent Application 2006/0170188, published 08/03/2006.

The recovery of energy during the braking of a car into compressed air is described by Guy Negre and Cyril Negre in United States Patent 6,363,723, issued 04/02/2002. The compressed air is then useful later in re-accelerating the vehicle with a pneumatic engine.

Summary of The Invention

Briefly, a magnetic air car embodiment of the present invention uses a magnetic motor to compress input air and save moderately compressed high-pressure air in storage tanks. The compressor and storage tanks deliver the high-pressure working air and operational flows to several stages of compressors that boost the pressures during driving to very high-pressure, then ultra high-pressure, then super high-pressure, and finally to extremely high-pressure.

A pneumatic torque converter uses jets of the extremely high-pressure to turn an input shaft of a transmission and differential. These, in turn, drive the powered wheels of a car. The compressors float a connecting shaft with matching vanes and impellers on opposite ends on air bearings to reduce shaft turning friction to near zero. The balance of forces between the two ends of a coupled turbo pair allows a simple air bearing design to operate safely and reliably at high rotational speeds. An alternative automobile embodiment of the present invention comprises a magnetic engine that drives an air compressor to fill on-board compressed air tanks. The magnetic engine is started initially by an electric starter motor and battery. Air bearings are used throughout to reduce friction to near zero in the compressors, compressor, magnetic engine, and air cycle engine. The compressed air tanks are filled to only moderate pressures, such as 200-PSI. Compressed air from the tanks and the air compressor is passed through air amplifiers to greatly increase the volume of input air. Amplified air is input to two compressors that are each ducted and driven to function as two-stage compressors. High-pressure jets from these compressors are applied to a pneumatic engine that converts it to rotating mechanical energy to drive the wheels of a car. Exhaust heat and compressed air are recycled by a recuperator to improve efficiency. These and other objects and advantages of the present invention will no doubt become obvious to those of ordinary skill in the art after having read the following detailed description of the preferred embodiments which are illustrated in the drawing figures.

In The Drawings

Fig. 1 is a function block diagram of a magnetic air car embodiment of the present invention; and

Fig. 2 is a flowchart diagram of a magnetic air car method embodiment of the present invention.

Detailed Description Of The Preferred Embodiments

Fig. 1 represents a magnetic air car embodiment of the present invention, and is referred to herein by the general reference numeral 100. The magnetic air car 100 includes a 12-VDC storage battery 102 to start a magnetic motor 104 by energizing the windings of an internal starter motor. The magnetic motor, once brought up to its starting speed, drives a star-rotor compressor 106. The star-rotor type compressors have rotors which are synchronized not to touch one another during operation. In one embodiment of the present invention, battery 102includes a sodium free complex silicon salt electrolyte as described in PCT published patent application WO 01/13454 A1, published 02/22/2001. Greensaver Technology Corporation (Ningbo, China) says they hold a patent for their so-called GREENSAVER BATTERY. Silicone Batteries USA imports these batteries to the US. See,www.siliconebatteriesusa.com/.

The silicone battery is marketed as not having most of the bad qualities of lead acid batteries, e.g., high internal resistance, poor cold temperature performance, and significant self discharging rates. Silicone batteries are said to be able to present more than 80% of their total capacity even at temperatures as low as +15°F. Filtered, ambient air or recycled pressurized air is pumped up to about 200-PSI by star-rotor compressor 106 to produce a high-pressure (HP) supply 108. A pair of tanks 110 and 112 are used to store pressurized air for release as HP supply 114 into a series of step-up compressors. A first coupled pair of these are compressors 120 and 122 which have a common shaft 124 floated on an air bearing 126. This combination may be referred to herein as primary turbo TWIN1. Similarly, a second pair of compressors 130 and 132 have a common shaft 134 floated on an air bearing 136. This combination may be referred to herein as primary turbo TWIN2. HP supply 114 drives a pelton-type impulse turbine side of each compressor 120, 122, 130, and 136, and the exhaust is released to atmosphere. A pressure multiplication is provided like in a turbofan jet engine on the driven side of each compressor 120, 122, 130, and 136. This produces a very high-pressure (VHP) supply 138 from HP supply 108. A large coupled pair of compressors 140 and 142 have a common rotating shaft 144 on an air bearing 146. This combination may be referred to herein as a secondary turbo BOOST. Compressors 140 and 142 are driven by HP supply 114. Compressor uses this to step up VHP supply 138 into an ultra high-pressure (UHP) supply 148. The driven sides of compressors 120 and 132 then step this up to a super high-pressure (SHP) supply 150 and 152. Both are applied to a laminar jet 154 to produce a laminar airflow 156 into the driven side of compressor 142. The final result of all the pressure step-ups through compressors 120, 122, 130, 132, 140, and 142, is extra high pressure (EHP) supply 160. This is applied to a pneumatic torque converter 162, the hydraulic equivalent of which is a standard automatic transmission torque converter used in automobiles. For example, this includes a pelton-type impulse turbine. Pneumatic torque converter 162 couples with a driveshaft to a transmission and differential 164. The output torque is then used to drive axles to wheels 166 and 168 of a car. Power throttling is provided by modulating HP supply 108 from the star-rotor compressor 106. Exhaust 169 from pneumatic torque converter 162 is ducted to a compressor pair 170 and 172. A shaft 174 on an air-bearing 176 couples these together. This combination may be referred to herein as exhaust turbo RECOVERY.

Ambient air drawn in by a filter 178, and recycle air from compressor 172, are input to compressor 170. They receive a boost that is applied to the input HP supply 5 114 through a priority valve 179 to boost acceleration while the car is under way. All ambient air exchange takes place through air filter 178. The compressor pair 170 and 172, as do the others, provides a multiplication in the compressive pressures in gases that pass through the vanes of the driven sides. The multiplication is on the order of 5x to 7x. An independent air bearing supply system includes a 12-VDC electric compressor 180, a dedicated air storage tank 182, and an air bearing supply pressure 184. A control system included with magnetic air car 100 must float all the air bearings 126, 136, 146, and 176 first, before allowing any supply pressure 108 or 114 to spool up any of the compressors. Any loss of air bearing supply pressure 184 must be immediately used to shut down supply pressure 108 and 114 to stop the compressors spinning. Air bearings could also be usefully employed in magnetic motor 104, starrotor compressor 106, and torque converter 162. The 12-VDC electric compressor 180 could be powered by battery 102. Accessories, other than electrically powered ones like power steering and power windows, can be provided with a mechanical power take off (PTO) from magnetic motor 104 or pneumatic torque converter 162. A small pneumatic motor could also be used to drive accessories like air conditioning, alternators, and generators from taps on the HP supply 108 or discharge from compressor 172. The compressors are put in pairs around respective air bearings 126, 136, 146, and 176 to balance the lateral forces applied to the vane ends of shafts 124, 134, 144, and 174. A proper balance eliminates Milankovitch-like wobbles, e.g., changes in the axial tilt, axial precession, and eccentricities of the turbo-shafts 124, 134, 146, and 176 over periods of time.

A particular type of oil-free air bearing used in connection with a turbocharger is reported by Minoru Ishimo, "Air Bearing for Automotive Turbocharger", in R&D Review of Toyota CDRL, Vol. 41, No. 3, (c) 2006, Toyota Central R&D Labs, Inc. Some of the details in the article may be useful in implementing compressors 120, 122, 130, 132, 140, 142, 170, and 172. Magnetic motor 104 could be implemented by consulting any number of articles, whitepapers, United States Patents and Patent Application on the subject, or purchased as a ready-to-install engine from several suppliers now or soon-to-be marketed such products. For example, see CYCCLONE MAGNETIC ENGINES developed in Queensland, Australia, and represented in the United States by Cycclone Magnetic Engines, Inc., Reno, NV (www.cycclone.us). And see, Perendev Magnetic Motor (Perendev-Power.com). For patents, see, MAGNETIC ENGINE, by Allan Limb, UNITED STATES PATENT APPLICATION 2005/0116567, published 6/02/2005. The MAGFORCE piston engine from Shinyeon Energy Research Center of Korea www.shinyeonenergy.com). June 13, 2007, Kedron Corporation (Fairview, TN) issued a news release, "Kedron Corporation Discovers a New Energy Source that is Extremely Inexpensive, Abundant and Pollution-free", which can be seen at www.kedroncorp.com/pressrelease.html. And see, "Harnessing Mechanical Energy From Strong Electromagnetic Forces Generated By The Spin Of Electrons", accessible on the Internet at www.kedroncorp.com/abstract.html.

In general, a magnetic air car uses a magnetic motor to compress input air and save moderately compressed high-pressure (HP) air in storage tanks. The compressor and storage tanks deliver the high-pressure working air and operational flows to several stages of compressors that boost the pressures during driving to very high-pressure (VHP), then ultra high-pressure (UHP), then super high-pressure (SHP), and finally to extremely high-pressure (EHP). A pneumatic torque converter uses jets of the EHP to turn an input shaft of a transmission and differential.

These, in turn, drive the powered wheels of a car. The compressors float a connecting shaft with matching vanes and impellers on opposite ends on air bearings to reduce shaft turning friction to near zero. The balance of forces between the two ends of a coupled turbo pair allows a simple air bearing design to operate safely and reliably at high rotational speeds. One theory of operation for a practical magnetic motor was offered at peswiki.com by: A magnet generates mechanical energy or does work when for example it pulls toward another magnet or a piece of metal. The powerful magnetic forces of two neodymium magnets can do much more work than simply pull themselves together over a distance. For example, welders put neodymium magnets to work to hold metal parts together for welding. However, the welder must also do work when pulling the magnet away from the metal. Many of us have contemplated the notion of putting permanent magnets to work to turn the wheels on a vehicle or to drive an electric generator without the addition of external energy. For example, if the welder could remove the magnet with little or no effort (work) then the magnet would have delivered a "net" amount of work. Imagine two powerful magnets pulling themselves together with great force. The work that is done as they pull themselves together could be used to turn an electric generator. However, not much work would be obtained from only one such event. To obtain more work in this manner the magnets must be pulled apart repeatedly so that they can continuously do work by repeatedly pulling themselves together. The amount of energy spent pulling them apart has to be significantly less than the amount derived when they come together thus leaving a useful net-yield of energy that is applied to turning the generator. Pulling two magnets apart along the same path they took to pull themselves together will of course require as much (or more) energy as the amount generated by the magnets when they come together. However, it has been discovered recently that two permanent magnets of a particular shape can be pulled apart along a prescribed path that requires less work compared to the amount of work produced when the magnets come together along a different path. This is possible because permanent magnets have at least one North and one South Pole, which gives polarity to their magnetic fields making the fields and the force in the field unevenly distributed. In an uneven field of magnetic force, it is not difficult to imagine different paths having different forces and thereby generating different amounts of work. Star-rotor compressor 106 can be like the fifth generation products marketed by StarRotor Corporation, Bryan, TX (See, starrotor.com). The Company reports that their compressor can process any vapor or gas with the only associated design consideration being the selection of materials compatible with the gases being compressed. The compressor works by using inner and outer star rotors, with seven and eight points respectively, that rotate on corresponding axes. A drive mechanism synchronizes the rotors so they do not bear on one another. Seals made with sacrificial coatings are used between the rotors and stationary porting components. As the rotors turn, a chamber enlarges, reaches a maximum volume, and then squeezes closed. Inlet gas enters through the intake port as the void opens. Once the gas is captured, the chamber volume is squeezed causing the pressure to increase. After a design pressure is reached, the gas pushes out through a discharge port. The chamber ports open eight times per rotation of an outer rotor, allowing the compressor to process large volumes of gas. The position of the leading edge of the discharge port determines the compression ratio. If the leading edge is positioned to make the discharge port large, the compression ratio will be small. If the leading edge is positioned to make the discharge port small, the compression ratio will be high. By using a sliding mechanism, the leading edge position can be changed on the fly, giving the compressor a variable compression ratio. A magnetic motor could be integrated within to drive the compressor.

In operation, an electric motor driving auxiliary compressor 180 immediately begins filling the air bearing tank 182 when an ignition key is turned to the run position. The air bearing tank 182 supplies the pressurized air needed to suspend the air bearing loads of each component, e.g., 40-PSI @ 3.8 cubic feet per minute (cfm). Pressure sensors detect when a predetermined minimum operating pressure is present, and the magnetic motor 104 and star-rotor compressor 106 are allowed to start-up. Auxiliary compressor 180 is cycled on-off by pressure controller switches to keep a constant supply of compressed air in the air bearing tank 182. When the car is not in use, the air bearings do not need to remain suspended. A timer is used to allow the air bearing equipped components to spin down. After enough inertia has been spent and the possibility of damage to the air bearings has been reduced to zero, the timer shuts-off air flow from the air bearing tank 180, and the car and all its engine components are stopped. Fig. 2 represents a magnetic air car method embodiment of the present invention, referred to herein by the general reference numeral 200. Method 200 includes a step 202 for charging a storage battery, a step 204 for using that storage battery to start a magnetic motor, and a step 206 for running that magnetic motor on-board a vehicle.

A step 208 draws in filtered air into a compressor. A step 210 provides a first stage of compression by driving the compressor with the magnetic motor. A step 212 stores the compressed air at an HP level, at about 200-PSI, in storage tanks on-board the vehicle. A portion of the compressed air at HP level is used in a step 213 to simultaneously power a totem-pole series of compressors. A step 214 provides multiple stages of air compression that begin with HP to VHP, then VHP to UHP, then UHP to SHP, and finally SHP to EHP levels. A step 216 uses EHP level compressed air in a pneumatic toque conversion to drive an input shaft to a transmission/differential. A step 218 provides gearing in that transmission, including several forward speeds, neutral, and reverse. A step 220 drives the vehicle's wheels to provide locomotion. Useful power is still available in an exhaust from the pneumatic toque conversion step 216, so a boost step 222 provides pressurized boost air to step 210 through a priority valve. Such boost may be needed for quick acceleration and passing other cars.

A power take-off (PTO) allows a step 224 to generate a battery charge, and a step 226 to run accessories like air conditioning. If air bearings are used, a step 240 independently produces an air bearing supply pressure, and a step 242 uses such pressure to float the bearings of the compressors before they are allowed to spool-up. Such air bearing supply pressure is maintained until well after spool-down of the compressors.

Referring again to Fig. 1, an extra boost can be had by changing compressor 172 to be a drive turbine type, instead of a compressor. This would facilitate high altitude applications, such as is needed in airplanes. An additional fifth turbocharger could be added to increase 5 high altitude operation even more. Although the present invention has been described in terms of the presently preferred embodiments, it is to be understood that the disclosure is not to be interpreted as limiting. Various alterations and modifications will no doubt become apparent to those skilled in the art after having read the above disclosure. Accordingly, it is intended that the appended claims be interpreted as covering all alterations and modifications as fall within the true spirit and scope of the invention.

What is claimed is:

In The Claims

1. A magnetic air car vehicle, comprising: a magnetic motor in which permanent magnets areused to turn an output shaft without using electromagnetic components; a first compressor connected to be driven by the magnetic motor and for producing compressed air at a high-pressure (HP) level, about 200-PSI; storage tanks on-board said vehicle for receiving compressed air at said HP level from said first compressor; a series of compressors ducted in totem-pole fashion to multi-stage compress said compressed air from the storage tanks and first compressor beginning with HP level to a very high-pressure (VHP) level, then from said VHP level to an ultra high-pressure (UHP) level, then from said UHP level to a super high-pressure (SHP), and finally from said SHP level to an extremely high-pressure (EHP) level of compressed air; a pneumatic torque converter for converting said EHP level of compressed air to drive an output shaft; a transmission/differential connected to receive said output shaft from the pneumatic torque converter, and that includes forward speeds, a neutral, and a reverse; and wheels connected to provide locomotion from the transmission/differential.
2. The vehicle of Claim 1, wherein each in said series of compressors includes air bearings.
3. The vehicle of Claim 1, wherein each in said series of compressors are arranged in pairs to share a common interconnecting shaft supported by air bearings.
4. The vehicle of Claim 1, wherein each in said series of compressors is driven in parallel with the others by a common source of compressed air at said HP level from the storage tanks and compressor.
5. The vehicle of Claim 1, wherein the first compressor is a star-rotor type in which the rotors are synchronized not to touch one another during operation.
6. A magnetic air car method, comprising: charging a storage battery; using said storage battery to start a magnetic motor; operating said magnetic motor on-board a vehicle; drawing in filtered air into a first compressor; providing a first stage of compression to said filtered air by driving said first compressor

with said magnetic motor; storing compressed air at a high-pressure (HP) level, at about 200-PSI, in storage tanks on-board said vehicle; using a portion of said compressed air at HP level to simultaneously power a totem-pole series of compressors; compressing said filtered air in multiple stages beginning with HP level to a very high-pressure (VHP) level, then VHP level to an ultra high-pressure (UHP) level, then UHP level to a super high-pressure (SHP), and finally SHP level to an extremely high-pressure (EHP) level of compressed air; pneumatic torque converting said EHP level of compressed air to drive an input shaft to a transmission/differential; gearing said transmission to include forward speeds, a neutral, and a reverse; and driving the wheels of said vehicle to provide locomotion.

7. The method of Claim 6, further comprising: providing a pressurized boost air through apriority valve from an exhaust from said pneumatic toque conversion to supplement said compressed air at HP level.
8. The method of Claim 6, further comprising: using a power take-off (PTO) to generate a battery charge and to run air conditioning and other accessories.

The method of Claim 6, further comprising: using air bearings in said compressors; and independently producing an air bearing supply pressure to float said air bearings before said compressors are allowed to spool-up; and maintaining said air bearing supply pressure until after a spool-down of said compressors.

Glossary

Air Bearings

Unlike contact-roller bearings, air bearings utilize a thin film of pressurized air to provide an exceedingly low friction load-bearing interface between surfaces. The two surfaces don't touch. Being non-contact, air bearings avoid the traditional bearing-related problems of friction, wear, particulates, and lubricant handling, and offer distinct advantages in precision positioning and high-speed applications. The fluid film of the bearing is air that flows through the bearing itself to the bearing surface. The design of the air bearing is such that, although the air constantly escapes from the bearing gap, the continual flow of pressurized air through the bearing is enough to support the working loads.

Air Bearing Turbochargers

The application of oil-free bearings such as air or magnetic types to automotive turbochargers is expected to realize a reduction in mechanical losses while eliminating oil consumption. As a result, engines fitted with oil-free turbochargers will offer an improved response and lower fuel consumption. A turbocharger with the improved bearing has been run at a rotational speed up to 200,000 rpm. Air Bearing Turbochargers are **creating a Turbomachinery Revolution.** Work is progressing on a revolution in the use of oil-free bearings that will eliminate the need for an oil-lubrication system in high-temperature, high-speed rotating machinery. The foremost challenge for this technology is the design of an oil-free turbine engine to power 21st century aircraft.

Air Cycle Machine

An air cycle machine for producing a supply of cooled air and including a compressor rotor fixed to a shaft which is driven by a turbine rotor, the turbine serving to expand air which is pressurized by the compressor. Magnetic bearings support the shaft radially in a housing for rotation about a precisely established axis while a magnetic thrust bearing keeps the shaft in a precisely fixed axial position. The housing and the bearings are constructed as two split sections to permit the housing and the bearings to be separated radially and facilitate repair and/or replacement of the shaft and other internal components.

Air Flow Amplifiers

An air flow amplifier of relatively high air flow amplification ratios in which a thin film of pressurized primary air flowing in a transverse direction is mechanically deflected to impinge on a generally frusto-conical surface tapering towards the throat of the amplifier. The deflecting action is produced by a deflection ring which is spaced inwardly from the amplifier's annular nozzle. The ring has an internal diameter substantially larger than the amplifier's throat so that secondary air entering through the ring may flow directly towards the frusto-conical surface to mix with the primary air flowing along that surface.

Ambient Air

Any unconfined portion of the atmosphere - open air, surrounding air. The outside air.

Outdoor air. The air immediately around us. Refers to untreated air existing in any location.

The air outside buildings and structures. This does not refer to indoor air, air in the workplace, or contaminated air discharged from a source. All space outside of buildings, stacks or exterior ducts. Open air or outdoor air. Ambient air is a blanket of gases surrounding the earth. At ground level, air is a mixture of invisible and odorless ... The portion of the atmosphere, external to buildings, to which the general public has access. Air, which surrounds the occupant or process in a space.

Armature Magnet

In electrical engineering, an armature generally refers to one of the two principal electrical components of an electromechanical machine - a motor or generator, but may also mean the pole piece of a permanent magnet or electromagnet, or the moving iron part of a solenoid or relay. The other component is the field winding or field magnet. The role of the "field" component is simply to create a magnetic field (magnetic flux) for the armature to interact with, so this component can comprise either permanent magnets, or electromagnets formed by a conducting coil. The armature, in contrast, must carry current so it is always a conductor or a conductive coil, oriented normal to both the field and to the direction of motion, torque (rotating machine), or force (linear machine). The armature's role is two-fold: (a) to carry current crossing the field, thus creating shaft torque (in a rotating machine) or force (in a linear machine), and (b) to generate an electromotive force ("EMF"). In the armature, an electromotive force ("EMF") is created by the relative motion of the armature and the field. When the machine is acting as a motor, this EMF opposes the armature current, and the armature converts electrical power to mechanical torque (and power, unless the machine is stalled) and transfers it to the load via the shaft. When the machine is acting as a generator, the armature EMF drives the armature current, and shaft mechanical power is converted to electrical power and transferred to the load. (In an induction generator, these distinctions are blurred, since the generated power is drawn from the stator, which would normally be considered the field.) A growler is used to check the armature for shorts, opens and grounds.

Asymmetry

Thermodynamics is asymmetrical in time: the entropy in a closed system can only increase with time. A consequence of this is Clausius' Second Law, which states that there is no thermodynamic process whose sole effect is to extract a quantity of heat from a colder reservoir and deliver it to a hotter reservoir. Particle physics Symmetry is one of the most powerful tools in particle physics, because it has become evident that practically all laws of nature originate in symmetries. Violations of symmetry therefore present theoretical and experimental puzzles that lead to a deeper understanding of nature. Asymmetries in experimental measurements also provide powerful handles that are often relatively free from background or systematic uncertainties. Parity violation *Main article: parity (physics)* Until the 1950s, it was believed that fundamental physics was left-right symmetric; i.e., that interactions were invariant under parity. Although parity is conserved in electromagnetism, strong interactions and gravity, it turns out to be violated in weak interactions. The Standard Model incorporates parity violation by expressing the weak interaction as a chiral gauge interaction. Only the left-handed components of particles and right-handed components of antiparticles participate in weak interactions in the Standard Model. A consequence of parity violation in particle physics is that neutrinos have only been observed as left-handed particles (and antineutrinos as right-handed particles). In 1956-1957 Chien-Shiung Wu, E. Ambler, R. W. Hayward, D. D. Hoppes, and R. P. Hudson found a clear violation of parity conservation in the beta decay of cobalt-60. Simultaneously, R. L. Garwin, Leon Lederman, and R. Weinrich modified an existing cyclotron experiment and immediately verified parity violation. CP violation *Main article: CP-violation* After the discovery of the violation of parity in 1956-57, it was believed that the combined symmetry of parity (P) and simultaneous charge conjugation (C), called *CP*, was preserved. For example, CP transforms a left-handed neutrino into a right-handed antineutrino. In 1964, however, James Cronin and Val Fitch provided clear evidence that CP symmetry was also violated in an experiment with neutral kaons. CP violation is one of the necessary conditions for the generation of a baryon asymmetry in the universe. Combining the CP symmetry with simultaneous time reversal (T) produces a combined symmetry called CPT symmetry. CPT symmetry must be preserved in any Lorentz invariant local quantum field theory with a Hermitian Hamiltonian. As of 2006, no violations of CPT symmetry have been observed. Baryon asymmetry of the universe *Main article: baryogenesis* The baryons (i.e., the protons and neutrons and the atoms that they comprise) observed in the universe are overwhelmingly matter as opposed to anti-matter. This asymmetry is called the baryon asymmetry of the universe. Isospin violation: Isospin is the symmetry transformation of the weak interactions. The concept was first introduced by Werner Heisenberg in nuclear physics based on the observations that the masses of the neutron and the proton are almost identical and that the strength of the strong interaction between any pair of nucleons is the same, independent of whether they are protons or neutrons. This symmetry arises at a more fundamental level as a symmetry between up-type and down-type quarks. Isospin symmetry in the strong interactions can be considered as a subset of a larger flavor symmetry group, in which the strong interactions are invariant under interchange of different types of quarks. Including the strange quark in this scheme gives rise to the Eight-fold Way scheme for classifying mesons and baryons. Isospin is violated by the fact that the masses of the up and down quarks are different, as well as by their different electric charges. Because this violation is only a small effect in most processes that involve the strong interactions, isospin symmetry remains a useful calculational tool, and its violation introduces corrections to the isospin-symmetric results.

[edit] In collider experiments Because the weak interactions violate parity, collider processes that can involve the weak interactions typically exhibit asymmetries in the distributions of the final-state particles. These asymmetries are typically sensitive to the *difference* in the interaction between particles and antiparticles, or between left-handed and right-handed particles. They can thus be used as a sensitive measurement of differences in interaction strength and/or to distinguish a small asymmetric signal from a large but symmetric background. A **forward-backward asymmetry** is defined as $A_{FB}=(N_F-N_B)/(N_F+N_B)$, where N_F is the number of events in which some particular final-state particle is moving "forward" with respect to some chosen direction (e.g., a final-state electron moving in the same direction as the initial-state electron beam in electron-positron collisions), while N_B is the number of events with the final-state particle moving "backward". Forward-backward asymmetries were used by the LEP experiments to measure the difference in the interaction strength of the Z boson between left-handed and right-handed fermions, which provides a precision measurement of the weak mixing angle.

A **left-right asymmetry** is defined as $A_{LR}=(N_L-N_R)/(N_L+N_R)$, where N_L is the number of events in which some initial- or final-state particle is left-polarized, while N_R is the corresponding number of right-polarized events. Left-right asymmetries in Z boson production and decay were measured at the Stanford

Linear Collider using the event rates obtained with left-polarized versus right-polarized initial electron beams. Left-right asymmetries can also be defined as asymmetries in the polarization of final-state particles whose polarizations can be measured; e.g., tau leptons.

A **charge asymmetry** or particle-antiparticle asymmetry is defined in a similar way. This type of asymmetry has been used to constrain the parton distribution functions of protons at the Tevatron from events in which a produced W boson decays to a charged lepton. The asymmetry between positively and negatively charged leptons as a function of the direction of the W boson relative to the proton beam provides information on the relative distributions of up and down quarks in the proton. Particle-antiparticle asymmetries are also used to extract measurements of CP violation from B meson and anti-B meson production at the BaBar and Belle experiments.

Back MMF Region

> In electrical systems, the term "back emf" actually refers to the equal and opposite force field accompanying the "forward emf", in any "symmetrical" electrical system. In magnetic systems, the corresponding term is "forward and back mmf", rather than "forward and back emf".

With the symmetrization of the EE model by Lorentz in 1892, just at the very beginning of electrical engineering, our electrical engineers have been trained to think, design, build, and deploy only symmetrical systems. Lorentz (at the bidding of J. P. Morgan) eliminated all remaining asymmetrical Maxwellian systems in the already sharply curtailed Heaviside equations, further limiting the already severely-limited Heaviside model and thus the electromagnetic science and technology that later grew up to be based on it. With the symmetrization of the Heaviside equations, those *arbitrarily* symmetrical equations became the present "electrical engineering" model, usually taught (falsely) as if they were Maxwell's original theory. Maxwell's theory contains 20 quaternion-like equations in 20 unknowns, and contains both symmetrical and asymmetrical systems. You can develop and build all sorts of asymmetrical systems with Maxwell's original theory, that cannot be built in accord with the symmetrized Heaviside theory now loosely referred to as "Maxwell's theory". All rotating motors actually turn themselves from the broken symmetry that is created inside them. The present engineers have been trained that they must pay to put extra energy into the system, just to break its symmetry. That of course is totally false. Otherwise, a rotating electron (with its continual spin) would not spin. To reiterate: In order to achieve self-rotation of a magnetic motor, one must produce broken symmetry (asymmetry) in that otherwise symmetrical system.

When we use laterally-symmetric bar magnets, which presently is essentially all that the industry makes, the field strength on the left side of each magnet is equal to its field strength on its right side. So when I have a stator N and a rotor S facing it, using such "laterally symmetrical forces", then in the forward mmf region (where the rotor S is approaching the stator N) the rotor pole is being *accelerated* by the attractive force from the stator rotor pole in the forward mmf region, and so is having *free angular momentum* being generated and stored in excess in the flywheel and shaft. That's real energy (specifically, energy x time). But when the rotor magnetic pole passes the stator magnetic pole, the direction of the mutual attraction force is now reversed (this is the *back mmf region*), and so the rotor is decelerating and thereby decelerating the previously accelerated shaft and flywheel. If the magnetic fields of each magnet are laterally symmetric, and nothing else is done, then in the back mmf region the system will freely take back (decelerate) all the excess energy freely stored in the accelerated flywheel and shaft in the previous forward mmf region. So the *net* force of this symmetrical arrangement is zero, and the *net* angular momentum generated and stored in the flywheel and shaft is also zero. In its back mmf (decelerating) region, the symmetrical permanent magnet motor system takes back all the free energy it stored in its earlier forward mmf (accelerating) region. So that silly thing will not give us any self-rotation and free energy, because of that lateral field strength symmetry of the bar magnets used as stator and rotor magnets. In a *normal* motor, we are trained to put in a coil (say, there in the back mmf region) and then *we pay* to put in a sudden surge of EM energy to that coil, so that it momentarily overrides (cancels) the back mmf force. In short, we momentarily make the system *asymmetrical*, so that its *net* back mmf is less than its *forward* mmf. That means that now the motor retains at least some of its excess acceleration and excess angular momentum added to the flywheel and shaft in its previous acceleration (forward mmf) zone, but *we are "paying" to have this occur*. Anyway, once that *broken symmetry* between forward and back mmfs is there, with the back mmf deliberately reduced to less than the forward mmf, the motor will self-rotate because of its own system asymmetry. And so we can add a "drag load" to the shaft, to soak up all that excess energy gained in each rotation, to do work to power the load. In that case, the motor rotates continuously, furnishing energy and power to power the load continuously. But it does not do it "for free", because we are paying for that broken symmetry all the time.

We are paying to break the symmetry, nothing else. Now let us reason together. Nature and this system do not care *how* we get that broken symmetry. If we wish to continually "pay" for it, and thus be tied to consuming fuel to get our "payment" energy for breaking the symmetry, we can do so. And that is precisely what everybody is and has been trained to do – ever since J. P. Morgan had Lorentz symmetrize the EE model itself, so our engineers would build only *symmetrical* systems. In that case, they will always have to *pay* to make the required broken symmetry, and that keeps the world firmly tied to its energy crisis and its escalating economical problems. But now suppose we contract with some nanocrystalline folks, to build a special "laterally asymmetrical" permanent magnet. They will lay down the plane of crystals for the side edge, with full magnetic field at the beginning layer (say, on the left), then lay an adjacent plane to the right but with a slightly weaker magnetic field, and they continue laying layers to the right with *successively weaker fields*. The result is a permanent magnet (bar magnet) with *asymmetric* field strengths laterally. Suppose we get two of those laterally asymmetric bar magnets, and use them with a rotor and stator and shaft to make a self-rotating motor. Suppose we mount the rotor (say, S-pole facing the stator) and stator (say, N-pole facing the rotor) so that, when the rotor S pole is rotating and approaching the stator N pole, the strong sides of both bar magnets are facing each other. This gives a certain acceleration added to the flywheel and shaft, and it stores up a certain amount of free angular momentum in the rotor and flywheel, while the rotor is traversing through that "forward mmf" zone. Then as the rotor magnetic

S pole passes the stator magnetic N pole and enters the back mmf zone of the system, the weaker sides of the magnets are now facing. So the *deceleration* in this back mmf zone is less than the *incoming acceleration* was (in the forward mmf zone). And that silly thing will sit there and self-rotate, till the end of time if nothing else bothers it or affects it, and so long as nothing breaks. And it will try to continually accelerate the flywheel and shaft with a *net acceleration* during each rotation.

This means I can now add a matched "drag (decelerating)" load to the accelerated shaft, and use all that extra stored energy from each rotational cycle, to freely power my load.

And if I match the load drag and the available "free asymmetry energy" net acceleration, that silly beast will sit there and self-rotate and power its load till the end of time, if nothing else intervenes. And now we do not have to pay anything for breaking the symmetry, after the initial costs of the laterally asymmetric magnets and assembling the system are paid!

Magnetic motors are powered by their broken symmetry, *not by what breaks the original symmetry*. We ourselves can continually pay to continually break the symmetry if we wish, or we can just build the system asymmetrically in the first place. Nature doesn't care. Note that, once the nanocrystalline folks have done their number to develop the laterally-asymmetry bar magnets, then such permanent magnets with laterally asymmetric field strengths can be put into production for not too much more than the cost of normal symmetric magnets. And at that point, anyone in the world can order some laterally-asymmetric permanent magnets, and easily assemble his own self-rotating motor and thus his own self-powering system.

Bearden Motor

Let us now turn to a simple and concrete example of a permanent magnet motor. In figure 1 we show the normal sort of idea the conventional engineer thinks of when confronted with the question of a self-energized permanent magnet motor.

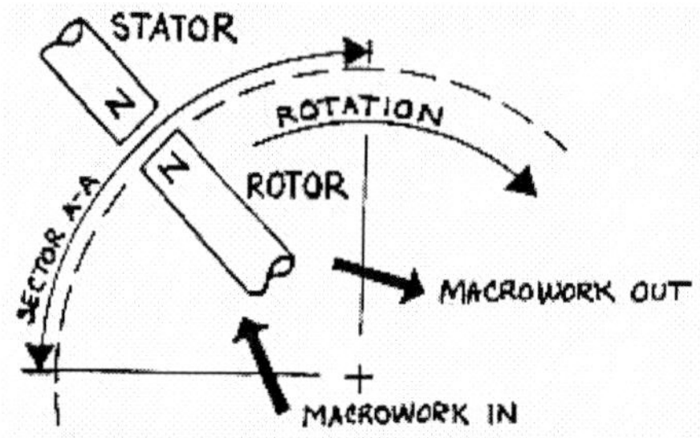

Figure 1. Permanent magnet motor with linear magnets and fields. We show parts of two permanent magnets: A fixed north pole as part of the rotor. We choose to examine the magnetic force field situation in a 90-degree sector A-A about the stator position show. We assume that the rotor is mounted in convenient fashion upon a shaft, with a flywheel to store and release energy. As can be seen, macrowork from the flywheel through the shaft must be furnished to drive the rotor during the first half of its travel through sector A-A. During the second half of the sector, work is accomplished by the rotor on the shaft, and is back through the shaft and stored in the flywheel as energy. In figure 2, sector A-A is spread out flat as a horizontal axis.

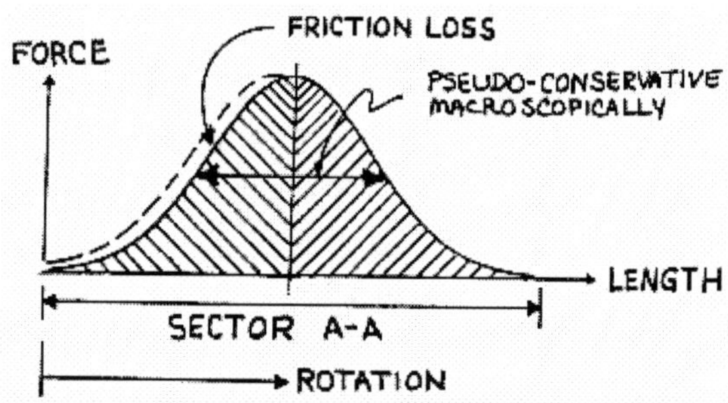

Figure 2. Macroscopic pseudo-conservation of energy. Rotation now appears from left to right along this line. On the vertical axis we show the magnitude of the repulsive force experienced by the rotor north pole N at each position along sector A-A. In the leftmost half of the diagram, the repulsive force between the two magnets requires that energy be expended from the flywheel through the shaft onto the rotor system. In the rightmost half, the repulsive force between the two magnets furnishes energy back to the rotor

system, and in turn through the shaft into the flywheel for storage. In each case, the energy involved is given by the area under that half of the curve. As can be seen, in the idealized system, as much work is required to be done on the rotor system to drive it to center (to the stator) as is performed back on the rotor system as it passes the stator and moves away to the right. In such a system with absolutely zero frictional losses, the best that one could do would be to break even, so that work in equals work out. Since any real system has at least a little friction, additional work (given by the area between the dotted line and the solid line in the left half of figure 2) is required to be put into the machine. Therefore the work out is always less than the work that must be put in, and the system is at best a "pump" of high efficiency, but of efficiency less than unity. Some hidden assumptions, such is the conventional wisdom. Now what are the hidden assumptions? First, one assumed symmetrical (we shall say, linear) fields and magnets. Second, one assumed a "closed macrosystem" with no interaction with teh vacuum/spacetime. That is, one assumed the electromagnetic model posed by classical electromagnetics, which is over 100 years old. A lot of physics has been discovered in the century that has passed since that model was formed! Third, one assumed that the "static force experienced by the fixed stator did not work. That is, again one was assuming the classical model. Let us further examine that assumption. That is true even when frictional loss is added, as shown by the inner dotted line in figure 4.

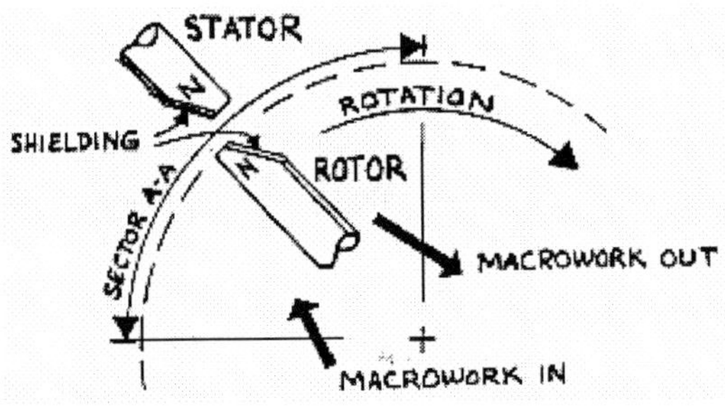

Figure 3. Permanent magnet motor with linear magnets and fields.

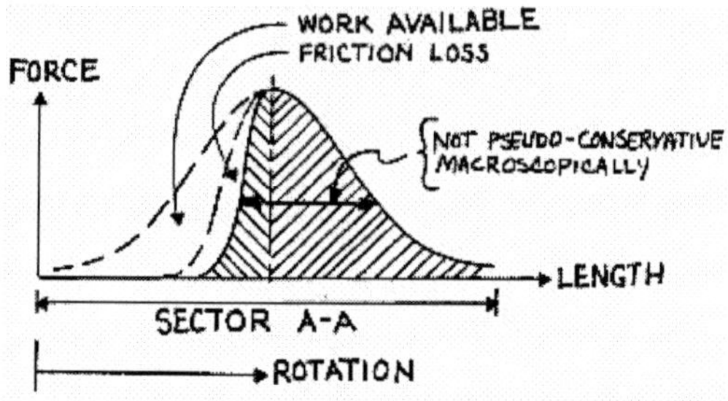

Figure 4. Macroscopic non-pseudoconservation of energy. As can be seen, now some remaining work is available to be taken out of the rotor/shaft/ flywheel system as <u>shaft horsepower</u>. In other words, this system can now work on another external system. Now refer to figures 5 and 6, which further clarify the situation and show the two microwork fluxes going on in each half of the rotation through sector A-A.

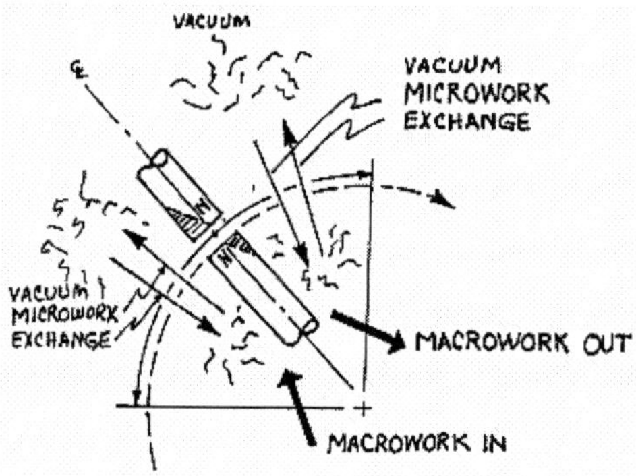

Figure 5. Permanent magnet motor: vacuum microwork exchange.

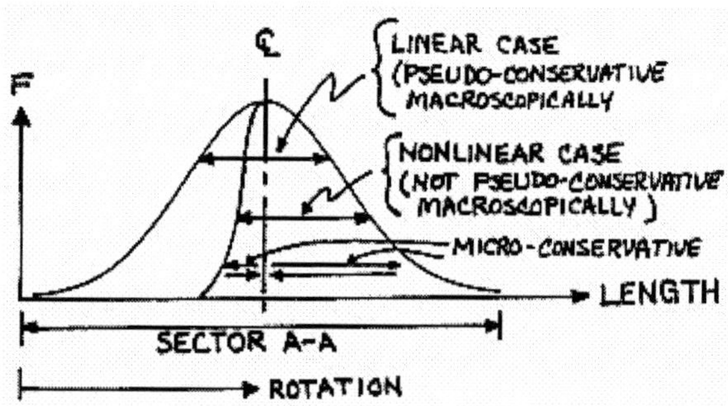

Figure 6. Pseudoconservation and non-pseudoconservation of energy. Note that the true (micro) conservation law is still obeyed -- even though the classical (pseudoconservation) law is clearly violated. That is, during each separate half of the situation, the magnitude of the microwork done on the rotor/vacuum system (and integrated into macrowork). But the work flux exchanged in the first half, because the fields and magnets are asymmetrical, compared half to half. Thus the asymmetrical system furnishes more integrated macrowork to the shaft and flywheel during the second half than it demands back during the first half. We have a "free energy" motor here after all, and we've just shown that a true self-energized permanent magnet motor can indeed be built. Indeed, any university or technical institute -- or researcher -- can fairly readily build a demonstration model, whether or not it's very practical. That almost no one does so, and almost no one "thinks the problem through" in modern physics terms rather than the old classical electromagnetism, is a commentary on how far we've gone at turning science into dogma.

BHmax

Maximum Energy Product, (BH) max: there is a point in the hysteresis loop where the product magnetizes the H force and the B induction reaching a maximum. The maximum value is called the Maximum Energy Product. This point is the magnet required value to program an energy. This parameter is usually used to describe how strong this permanent magnetic material is. Its unity is Gauss Oersted. One MGOe means 1.000.000 Gauss Oersted.

Broken Symmetry

Broken symmetry is a concept, developed by Lee and Yang, used in mathematics and physics when an object breaks either rotational symmetry or translational symmetry. That is, when one can only rotate an object in certain angles or when one is able to discern if the object has been shifted sideways (unless one shifts by a whole number of lattice units). For example, imagine a jumping bean sitting atop a Mexican hat like the one pictured here. It is in a rotationally symmetric state until it inevitably hops and falls down to a lower equilibrium, breaking the rotational symmetry.

Coanda Effect

The Coandă effect (IPA: [ˈkwandə]) is the tendency of a fluid jet to stay attached to an adjacent curved surface that is very well shaped. The principle was named after Romanian Henri Coandă, who was the first to recognize the practical application of the phenomenon in aircraft development.

Coil

An electromagnetic coil (or simply a "coil") is formed when a conductor (usually a solid copper wire) is wound around a core or form to create an inductor or electromagnet. One loop of wire is usually referred to as a turn, and a coil consists of one or more turns. For use in an electronic circuit, electrical connection terminals called taps are often connected to a coil. Coils are often coated with varnish and/or wrapped with insulating tape to provide additional insulation and secure them in place. A completed coil assembly with taps etc. is often called a winding. A transformer is an electromagnetic device that has a primary winding and a secondary winding that transfers energy from one electrical circuit to another by magnetic coupling without moving parts. The term tickler coil usually refers to a third coil placed in relation to a primary coil and secondary coil. A coil tap is a wiring feature found on some electrical transformers, inductors and coil pickups, all of which are sets of wire coils. The coil tap(s) are points in a wire coil where a conductive patch has been exposed (usually on a loop of wire that extends out of the main coil body). As self induction is larger for larger coil diameter the current in a thick wire tries to flow on the inside. The ideal use of copper is achieved by foils. Sometimes this means that a spiral is a better alternative. Multilayer coils have the problem of interlayer capacitance, so when multiple layers are needed the shape needs to be radically changed to a short coil with many layers so that the voltage between consecutive layers is smaller (making them more spiral like).

Composite Foam Grid Batteries

Firefly's technology is an innovative material science that removes almost all limitations of current lead-acid battery products. The materials also hold the promise of major simplification for manufacturing of lead-acid batteries and will potentially deliver more flexible form factors or configurations, which may be the catalyst to change the entire distribution and profitability models of the battery industry. As he began his research into lead acid battery chemistry and structure, Kurt Kelley discovered that much of the lead in the grid structure of conventional batteries can be replaced by a totally new type of grid material. Of course, once the basic material was determined to have the requisite physical and chemical properties, much subsequent research and testing was required to determine the optimum configuration and "architecture" within the battery itself. These results are confirmed in recently granted U.S. Patents. In the advanced battery architectures that Firefly has perfected, the MicrocellTM composite foam "grids" are impregnated with a slurry of lead oxides which are then formed up to the sponge lead and lead dioxide in the normal fashion. Because of the foam structure, the resultant negative and positive plates have enormous surface-area advantages over conventional lead acid grid structures. This results in much-improved active material utilization levels (i.e. from the historical 20-50% up into the range of 70-90%) as well as enhanced fast-recharge capability and greater high-rate / low-temperature discharge times.

Compressed Air Cars

Compressed air cars are powered by engines fueled by compressed air, which is stored in a tank at high pressure such as 30 MPa (4500 psi or 300 bar). The storage tank is likely to be made of carbon-fiber in order to reduce its weight while achieving the necessary strength. Instead of mixing fuel with air and burning it to drive pistons with hot expanding gases; *compressed air cars* use the expansion of compressed air to drive their pistons. The idea is not new. There have been prototype cars since the 1920s and compressed air has been used in torpedo propulsion as well. Compressed air is a heavy way of storing fuel,

300 litres (11 cu ft) air at 30 MPa (4,500 psi) contains about 16 kWh of energy (the equivalent of 1.7 liters [0.44 US gal, 0.37 imp gal] of gasoline, assuming a 100% efficiency of the engine). During rupture testing, the tank cracks, but does not break up, producing no splinters or fragments. All four major manufacturers that are developing air cars have designed safety features into their containers. In contrast to hydrogen's issues of damage and danger involved in high-impact crashes, air, on its own, is non-flammable. It was reported on Discovery's *Beyond Tomorrow* that on its own, carbon-fiber is brittle and can split under sufficient stress, but creates no shrapnel when it does so. Carbon-fiber tanks safely hold air at a pressure somewhere around 4500psi, which is good when compared with steel tanks. Emission output: Compressed air cars are emission-free at the 'tailpipe'. Since a compressed air car's source of energy is usually electricity, its total environmental impact depends on how clean the source of this electricity is. Due to this reason, a compressed air car's emission output can vary both with location and time. Different regions can have very different sources of power, ranging from high-emission power sources such as coal to zero-emission power sources such as wind. A given region can also update its electrical power sources with time, thereby improving or worsening emissions output.

Conductor Wire

An **electromagnetic coil** (or simply a "coil") is formed when a conductor (usually a solid copper wire) is wound around a core or form to create an inductor or electromagnet. One loop of wire is usually referred to as a *turn*, and a coil consists of one or more turns. For use in an electronic circuit, electrical connection terminals called taps are often connected to a coil. Coils are often coated with varnish and/or wrapped with insulating tape to provide additional insulation and secure them in place. A completed coil assembly with taps etc. is often called a *winding*. A transformer is an electromagnetic device that has a *primary winding* and a *secondary winding* that transfers energy from one electrical circuit to another by magnetic coupling without moving parts. The term *tickler coil* usually refers to a third coil placed in relation to a primary coil and secondary coil. A coil tap is a wiring feature found on some electrical transformers, inductors and coil pickups, all of which are sets of wire coils. The coil tap(s) are points in a wire coil where a conductive patch has been exposed (usually on a loop of wire that extends out of the main coil body). As self induction is larger for larger coil diameter the current in a thick wire tries to flow on the inside. The ideal use of copper is achieved by foils. Sometimes this means that a spiral is a better alternative. Multilayer coils have the problem of interlayer capacitance, so when multiple layers are needed the shape needs to be radically changed to a short coil with many layers so that the voltage between consecutive layers is smaller (making them more spiral like).

Dynamical Theory

The dynamical theory of diffraction considers the wave field in the periodic potential of the crystal and takes into account all multiple scattering effects. Unlike the kinematic theory of diffraction which describes the approximate position of Bragg or Laue diffraction peaks in reciprocal space, dynamical theory corrects for refraction, shape and width of the peaks, extinction and interference effects. Graphical representations are described in dispersion surfaces around reciprocal lattice points which fulfill the boundary conditions at the crystal interface.

Electrical Energy

Electric energy is the potential energy associated with the conservative Coulomb forces between charged particles contained within a system, where the reference potential energy is usually chosen to be zero for particles at infinite separation. It can be defined as the amount of work one must apply to (massless) charged particles to bring them from infinite separation to some finite proximity configuration. This is also equal to the negative of the work of the Coulomb forces that the particles exert on each other during the quasistatic move.

$$U_E = W_{app} = -W_\infty$$

Where W_{app} is the work required to bring the system to a certain finite proximity configuration. *"app"* stands for applied, because this is work that must be applied to the system (or be supplied by another form of energy contained by the system) to configure it W_{∞} is the work done by electrostatic inter-particle Coulomb forces during the move from infinity. Sometimes people refer to the potential energy of a charge in an electric field. This actually refers to the potential energy of the system containing the charge and the *other* particles that created the electric field. Furthermore, to calculate the work required to bring a charged particle into the vicinity of other particles, it is sufficient to know only the field generated by the other particles and the charge of the particle being moved. The field of the moving particle and the individual charges of the other particles do not need to be known. Finally, it must be stressed that, even though this article talks about moving particles, the Coulomb force law on which this discussion is based only holds in the case of electrostatic systems. Therefore, any movement would have to be a quasistatic process.

Electric Motor

An **electric motor** uses electrical energy to produce mechanical energy. The reverse process, that of using mechanical energy to produce electrical energy, is accomplished by a generator or dynamo. Traction motors used on locomotives and some electric and hybrid automobiles often perform both tasks if the vehicle is equipped with dynamic brakes. Electric motors are found in household appliances such as fans, refrigerators, washing machines, pool pumps, floor vacuums, clothes dryers, hair dryers, forced air home furnaces, computers requiring cooling, many DVR recorders, attic fans, range hoods, bathroom fans, boot dryers, pellet stoves, many electric heaters, heat guns, and, and fan-forced ovens. They are also found in many other devices such as computer equipment, in its disk drives, printers, and fans; and in some sound and video playing and recording equipment as DVD/CD players and recorders, tape players and recorders, and record players. Electric motors are also found in several kinds of toys such as some kinds of vehicles, fans, cool water mist fans, blow torches, pencil sharpeners, fan-forced ovens, and, and robotic toys.

Electromagnetic Energy (heat)

Electromagnetic radiation (sometimes abbreviated **EMR**) takes the form of self-propagating waves in a vacuum or in matter. EM radiation has an electric and magnetic field component which oscillate in phase perpendicular to each other and to the direction of energy propagation. Electromagnetic radiation is classified into types according to the frequency of the wave, these types include (in order of increasing frequency): radio waves, microwaves, terahertz radiation, infrared radiation, visible light, ultraviolet radiation, X-rays and gamma rays. Of these, radio waves have the longest wavelengths and Gamma rays have the shortest. A small window of frequencies, called visible spectrum or light, is sensed by the eye of various organisms, with variations of the limits of this narrow spectrum. EM radiation carries energy and momentum, which may be imparted when it interacts with matter.

EM Energy (real observable photons)

Is it indeed possible to collect some of the electromagnetic energy coming off of our planet as it hurls through the solar system in this bubble traveling thru deep space? Ah ha, said NASA Scientists, well there is all the power we need the physicists were right in their calculations, well actually they underestimated this massive power. So let's build a giant satellite with 20-mile long rotating energy tails that spin to collect the energy from the Earths field for powering up satellites, ISS, Hubble or Moon Base, Low Orbit Space Hotel balloon? Refill stations? We can send the energy wireless to wherever we need it and use a space laser to get it there? It makes a good Near Earth Object Defense System and unfortunately $2 Billion in cost is a lot, but well worth it said one Los Alamos Scientist. You see Free Energy is available we just need to build the proper infrastructure to collect it.

Energy

Energy (from the Greek ἐνέργεια - *energeia*, "activity, operation", from ἐνεργός - *energos*, "active, working is a scalar physical quantity that is an attribute of objects and systems that is conserved in nature. In physics textbooks, energy is often defined as the ability to do work or to cause change. Several different forms of energy exist to explain all known natural phenomena. These forms include (but are not limited to) kinetic, potential, thermal, gravitational, sound, light, elastic, and electromagnetic energy. Any form of energy can be transformed into another form, but the total energy always remains the same. This principle, the conservation of energy, was first postulated in the early 19th century, and applies to any isolated system. According to Noether's theorem, the conservation of energy is a consequence of the fact that the laws of physics do not change over time. Although the total energy of a system does not change with time, its value may depend on the frame of reference. For example, a seated passenger in a moving airplane has zero kinetic energy relative to the airplane, but non-zero kinetic energy relative to the Earth.

Entropy

In many branches of science, **entropy** is a measure of the disorder of a system. The concept of entropy is particularly notable as it is applied across physics, information theory and mathematics. In thermodynamics (a branch of physics), entropy, symbolized by *S*, is a measure of the unavailability of a system's energy to do work. It is a measure of the disorder of molecules in a system, and is central to the second law of thermodynamics and to the fundamental thermodynamic relation, both of which deal with physical processes and whether they occur unexpectedly. Spontaneous changes in isolated systems occur with an increase in entropy. Unexpected changes tend to average out differences in temperature, pressure, density, and chemical potential that may exist in a system, and entropy is thus a measure of how great the unexpected changes are. The word "entropy" is derived from the Greek *εντροπία* "a turning towards" (*εν-* "in" + *τροπή* "a turning").

Extremely High Pressure (EHP)

A magnetic air car vehicle, comprising: a magnetic motor in which permanent magnets are used to turn an output shaft without using 5 electromagnetic components; a first compressor connected to be driven by the magnetic motor and for producing compressed air at a high-pressure (HP) level, about 200-PSI; storage tanks on-board said vehicle for receiving compressed air at said HP level from said first compressor; a series of compressors ducted in totem-pole fashion to multi-stage compress said compressed air from the storage tanks and first compressor beginning with HP level to a very high-pressure (VHP) level, then from said VHP level to an ultra high-pressure (UHP) level, then from said UHP level to a super high-pressure (SHP), and finally from said SHP level to an **extremely high-pressure (EHP) level** of compressed air; a pneumatic torque converter for converting said EHP level of compressed air to drive an output shaft; a transmission/differential connected to receive said output shaft from the pneumatic torque converter, and that includes forward speeds, a neutral, and a reverse; and wheels connected to provide locomotion from the transmission/differential.

Faraday's Law of Induction

The law was discovered by Michael Faraday in 1831 and independently at the same time by Joseph Henry. Quantitatively, the law takes the following form:

$$\mathcal{E} = -\frac{d\Phi_B}{dt}.$$

Where $\mathcal{E}$ is the electromotive force (EMF) in volts Φ_B is the magnetic flux through the circuit (in webers). The direction of the electromotive force (the negative sign in the above formula) is given by Lenz's law. The meaning of "flux through the circuit" is elaborated upon in the examples below. Traditionally, two different ways of changing the flux through a circuit are recognized. In the case of **transformer EMF** the idea is to alter the field itself, for example by changing the current originating the field (as in a transformer). In the case of **motional EMF**, the idea is to move all or part of the circuit through the magnetic field, for example, as in a homopolar generator.

Ferromagnetism

Ferromagnetism is the basic mechanism by which certain materials (such as iron) form permanent magnets and/or exhibit strong interactions with magnets; it is responsible for most phenomena of magnetism encountered in everyday life (for example, refrigerator magnets). The attraction between a magnet and ferromagnetic material is "the quality of magnetism first apparent to the ancient world, and to us today," according to a classic text on ferromagnetism.[1]

All permanent magnets (materials that can be magnetized by an external magnetic field and which *remain* magnetized after the external field is removed) are either ferromagnetic or ferrimagnetic, as are the metals that are noticeably attracted to them. Historically, the term **ferromagnet** was used for any material that could exhibit spontaneous magnetization: a net magnetic moment in the absence of an external magnetic field. This general definition is still in common use. More recently, however, different classes of spontaneous magnetization have been identified when there is more than one magnetic ion per primitive cell of the material, leading to a stricter definition of "ferromagnetism" that is often used to distinguish it from ferrimagnetism. In particular, a material is "ferromagnetic" in this narrower sense only if *all* of its magnetic ions add a positive contribution to the net magnetization. If some of the magnetic ions *subtract* from the net magnetization (if they are partially *anti*-aligned), then the material is "ferrimagnetic". If the ions anti-align completely so as to have zero net magnetization, despite the magnetic ordering, then it is an antiferromagnet. All of these alignment effects only occur at temperatures below a certain critical temperature, called the Curie temperature (for ferromagnets and ferrimagnets) or the Néel temperature (for antiferromagnets).

Free Electromagnetic Energy Wind from Active Vacuum

Vacuum" is what we usually think of as the empty space left after all the air is removed. However, in modern physics it is well-known that space is not such an "emptiness" at all. Instead, it is filled with energetic particles that appear and disappear with extraordinary speed. Hence "the vacuum" in physics is more like a seething cauldron, boiling fiercely {1}. The energy density of this boiling is so great that it literally boggles one's mind {2}. A bit of "empty space" the size of a sugar cube contains enough "seething electromagnetic energy" to power all the electrical loads on earth for millions of years. Obviously, if we can extract just a tiny fraction of this energy in electromagnetic form, we can use it to power our loads for free or nearly so. We need only power the "gating" or "switching" of the flowing energy, and thus cause a great deal of work to be performed, even though we ourselves input little or no energy.

Can the EM Energy of the Vacuum be "Tapped" and Extracted for our Use?

It's widely believed that it is impractical to try to extract very much usable energy from the seething, highly energetic vacuum. Oh yes, the tiny little Lamb shift {3} of the orbit of one electron in a certain atom is admittedly due to vacuum energy interaction. Well, although the Lamb shift is a very small effect of a single electron in a single atom, the *energy density* of that interaction is greater than the surface energy density of the sun!

And, oh yes, the Casimir effect {4} — a vacuum-induced attraction of two conducting plates when separated but close — also is due to vacuum energy interaction, but it too is a very small effect and not practical to consider. Then the assertion is often made — particularly by electrical power engineers, who normally are not well-acquainted with modern physics — that thermodynamics prohibits extracting and using vacuum energy. That is quite untrue, as we have explained, and as Cole and Puthoff have rigorously demonstrated. *Thermodynamics does not explain how to do it practically, but it does permit it technically.* The principle of superposition of charge and the field effects of charges, ubiquitously assumed and verified in countless classical experiments, also tells us a magnificent thing: *The source charge production of its associated fields and potentials is additive and coherent as the source charge is increase . Within reason, doubling or tripling the amount of energy-collecting and transducing charged particles (source charges) at a given point will give double or triple the resulting potentials and fields (gathered EM energy and local EM energy winds) produced by that source charge collection, and so on. So the proven broken symmetry of opposite charge collections — such as on the ends of a dipole or dipolarity — need not be a small effect at all! We can use as many unit charges as we wish in each one of those "separated opposite charges" in a dipole, so long as we hold the charges apart with a proper restraining dielectric.* In a normal electrical circuit (as between the terminals of a generator or battery), we can have many positive and negative charges involved, so that the EM energy extracted from the vacuum can be very powerful. In other words, the dipole can readily be made essentially as powerful as we need for our usual specific purposes. We can easily produce a megawatt or even 1,000 megawatts of energy flow per second, by making a bigger dipole consisting of more separated charges. *All we have to do is pay to make the dipole once. Then leave it alone and do not allow it to be destroyed* by the inane way that engineers normally build their circuits and power their loads. The latter problem — which is the only real problem preventing the engineers from easily producing all the cheap, clean electrical power they wish, without burning fuel, consuming fuel rods, building pipelines, building dams, etc. — is not being furiously funded and worked on as a national Manhattan Project. Indeed, the very notion continues to be castigated and ridiculed by the "defenders of the faith". And that is perhaps the greatest scientific travesty of our times. It simply reveals the "soft underbelly" of our present energy science, which does not even include what powers its circuits. Every nation must soon get cheap clean electrical energy from the vacuum and develop self-powering electrical power systems quickly or no nation will ever be safe again. Radicals and terrorists will see to it, if we lose the energy war. So COP>1.0 electrical power systems have become both a moral imperative and a national imperative, in my opinion. Either we develop them and establish them worldwide or we perish, as simple as that. If enough of us determine to do it, and if we hang together and get a little funding — and if we have a little luck— hopefully it will be accomplished.

Free Energy

In short, free energy is that portion of any First-Law energy that is available for doing thermodynamic work; *i.e.*, work mediated by thermal energy. Since free energy is subject to irreversible loss in the course of such work and First-Law energy is always conserved, it is evident that free energy is an expendable, Second-Law kind of energy that can make things happen within finite amounts of time. The free energy functions are Legendre transforms of the internal energy. For processes involving a system at constant pressure p and temperature T, the Gibbs free energy is the most useful because, in addition to subsuming any entropy change due merely to heat flux, it does the same for the pdV work needed to "make space for additional molecules" produced by various processes. (Hence its utility to solution-phase chemists, including biochemists.) The Helmholtz free energy has a special theoretical importance since it is proportional to the logarithm of the partition function for the canonical ensemble in statistical mechanics. (Hence its utility to physicists; and to gas-phase chemists and engineers, who do not want to ignore pdV work.) The (historically earlier) Helmholtz free energy is defined as $A = U - TS$, where U is the internal energy, T is the absolute temperature, and S is the entropy. Its change is equal to the amount of reversible work done on, or obtainable from, a system at constant T. Thus its appellation "work content", and the designation A from *Arbeit*, the German word for work. Since it makes no reference to any quantities involved in work (such as p and V), the Helmholtz function is completely general: its decrease is the maximum amount of work which can be done *by* a system, and it can increase at most by the amount of work done *on* a system.

Fuel Cells

A **fuel cell** is an electrochemical conversion device. It produces electricity from fuel (on the anode side) and an oxidant (on the cathode side), which react in the presence of an electrolyte. The reactants flow into the cell, and the reaction products flow out of it, while the electrolyte remains within it. Fuel cells can operate virtually continuously as long as the necessary flows are maintained. Fuel cells are different from electrochemical cell batteries in that they consume reactant, which must be replenished, whereas batteries store electrical energy chemically in a closed system. Additionally, while the electrodes within a battery react and change as a battery is charged or discharged, a fuel cell's electrodes are catalytic and relatively stable. Many combinations of fuel and oxidant are possible. A hydrogen cell uses hydrogen as fuel and oxygen (usually from air) as oxidant. Other fuels include hydrocarbons and alcohols. Other oxidants include chlorine and chlorine dioxide.

Gatekeepers

A **gatekeeper** is defined as someone who controls access to something. It also refers to individuals who decide whether a given message will be distributed by a mass medium. The gatekeepers, who represent the envious industries, the government, universities, scientists and critical journalists. They are the major cause of all travail on this earth. They have no imagination and do not believe in anything. They only believe in what can physically be observed in the present and have no vision for our future. I have clearly witnessed that many so-called scientific intellectuals in high positions of influence are FRAUDS. Why? Because they have the mentality of a parrot. They have simply memorized the great contributions of important scientific innovators of the past and can quote those contributions like a parrot. Yet at the same time, they have absolutely no true understanding or **curiosity** regarding the subjects they quote.

Gauss's law

In physics, **Gauss's law**, also known as **Gauss's flux theorem**, is a law relating the distribution of electric charge to the resulting electric field. Gauss's law states that:

The electric flux through any closed surface is
proportional to the enclosed electric charge.

In integral form (see below section for details) the equation is:

$$\oint_S \mathbf{E} \cdot d\mathbf{A} = \frac{Q_{enclosed}}{\varepsilon_0}$$

Gauss' law for magnetism

In physics, **Gauss's law for magnetism** is one of Maxwell's equations, the four equations that underlie classical electrodynamics. It states that the magnetic field **B** has divergence equal to zero, in other words, that it is a solenoidal vector field. It is equivalent to the statement that magnetic monopoles do not exist. Rather than "magnetic charges", the basic entity for magnetism is the magnetic dipole. (Of course, if monopoles were ever found, the law would have to be modified, as elaborated below.) Gauss's law for magnetism can be written in two forms, a *differential form* and an *integral form*. These forms are equivalent due to the divergence theorem.

Note that the terminology "Gauss's law for magnetism" is widely used, but not universal. The law is also called "Absence of free magnetic poles"(or some variant); one reference even explicitly says the law has "no name".

Generator

In electricity generation, an **electrical generator** is a device that converts mechanical energy to electrical energy, generally using electromagnetic induction. The reverse conversion of electrical energy into mechanical energy is done by a motor, motors and generators have many similarities. A generator forces electric charges to move through an external electrical circuit, but it does not *create* electricity or charge, which is already present in the wire of its windings. It is somewhat analogous to a water pump, which creates a flow of water but does not create the water inside. The source of mechanical energy may be a reciprocating or turbine steam engine, water falling through a turbine or waterwheel, an internal combustion engine, a wind turbine, a hand crank, the sun or solar energy, compressed air or any other source of mechanical energy.

Gyroscopic Particles

"Gyroscopic Particles influence one another and subsequently, SUCH INFINITELY-POSSIBLE-DEGREE-GYRATIONS FORM INFINITE TYPES OF MATTER." The scientists at the Stanford Linear Accelerator Center had light beams collide with one another. The GYROSCOPIC PARTICLE composition of light causes such gyroscopic particles to mechanically influence one another and, as a result, move at right angles to the force of one another in collision. As a result, such gyroscopic particles began gyrating around one another to form matter.

Gyroscopic Type MASSERGIES

"....Heat is electromagnetic energy (consisting of gyroscopic massergies*). Gyroscopic massergies* (or electromagnetic energy) comprise all Matter. Alterations in the heat (gyroscopic massergies*) of Matter cause a change in the amount of (gyroscopic massergies*) of Matter in accordance with E=mc^2."

Horse Power (HP)

Horsepower (**hp** or **HP**) is the name of several non-metric units of power. In scientific discourse, the term "horsepower" is rarely used because of its various definitions and the already existent SI unit for power, the watt (W). However, use of the term "horsepower" persists as a legacy in many languages and industries, particularly as a unit of measurement of the maximum power output of internal-combustion engines of automobiles; and often of trucks, buses, and ships. The use of 'HP' is being slowly replaced by kW (kilowatt) and MW (megawatt).

There are two important factors to consider when evaluating the measurement of "horsepower":

The inconsistent *definitions* of the "horsepower" unit itself The various *standards* used in measuring the value of "horsepower" These factors can be combined in unexpected ways — the power output for an engine rated at "100 horsepower" might not be what a reader expects. For this reason, various groups have attempted to standardize not only the definition of "horsepower" but the measurement of "horsepower". In the interim, more confusion may surface.

Horse Pressure

A magnetic air car vehicle, comprising: a magnetic motor in which permanent magnets are used to turn an output shaft without using 5 electromagnetic components; a first compressor connected to be driven by the magnetic motor and for producing compressed air at a high-pressure (HP) level, about 200-PSI; storage tanks on-board said vehicle for receiving compressed air at said HP level from said first compressor; a series of compressors ducted in totem-pole fashion to multi-stage compress said compressed air from the storage tanks and first compressor beginning with HP level to a very high-pressure (VHP) level, then from said VHP level to an ultra high-pressure (UHP) level, then from said UHP level to a super high-pressure (SHP), and finally from said SHP level to an **extremely high-pressure (EHP) level** of compressed air; a pneumatic torque converter for converting said EHP level of compressed air to drive an output shaft; a transmission/differential connected to receive said output shaft from the pneumatic torque converter, and that includes forward speeds, a neutral, and a reverse; and wheels connected to provide locomotion from the transmission/differential. This build up of pressure is called Horse Pressure™.

Hybrids

A **hybrid** is the combination of two or more different things, aimed at achieving a particular objective or goal. A **hybrid electric vehicle** (HEV) is a hybrid vehicle which combines a conventional propulsion system with a rechargeable energy storage system (RESS) to achieve better fuel economy than a conventional vehicle. It includes a propulsion system additional to the electric motors, to be not hampered by range from a charging unit like a battery electric vehicle (BEV). Modern mass-produced HEVs prolong the charge on their batteries by capturing kinetic energy via regenerative braking, and some HEVs can use the internal combustion engine (ICE) to generate electricity by spinning an electrical generator (often a motor-generator) to either recharge the battery or directly feed power to an electric motor that drives the vehicle. Many HEVs reduce idle emissions by shutting down the ICE at idle and restarting it when needed (start-stop system). An HEV's engine is smaller than a non-hybrid petroleum fuel vehicle and may be run at various speeds, providing more efficiency. HEVs became widely available to the public in the late 1990s with the introduction of the Honda Insight and Toyota Prius. HEVs are viewed by some automakers as a core segment of the future automotive market. *Futurist* magazine recently included hybrid electric vehicles as cars of the near future.

ISO Thermo Expansion

One of the biggest problems facing the magnetic engine and air flow design industry today has to deal with heat and friction. We believe that we may have the solution to this problem. We believe that by utilizing lightweight materials along with air bearing design modifications, and with the unique design of the air flow device itself, that we may have eliminated the heat and friction problem associated with magnetic engines and air flow devices. We call this methodology "**ISO Thermo Expansion Turbine Design**". By eliminating heat and friction from our design, we believe that magnetic engines and air flow devices may now be feasible for generating electricity and operating a transmission at a much lower cost than traditional methods.

Joules

The **joule** is the derived unit of energy in the International System of Units. It is defined as:

$$1\,\mathrm{J} = 1\,\mathrm{kg}\cdot\frac{\mathrm{m}^2}{\mathrm{s}^2}$$

One joule is the amount of energy required to perform the following actions:

The work done by a force of one newton traveling through a distance of one meter;

The work required to move an electric charge of one coulomb through an electrical potential difference of one volt; or one coulomb volt, with the symbol C·V; The work done to produce power of one watt continuously for one second; or one watt second (compare kilowatt hour), with the symbol W·s. Thus a kilowatt hour is 3,600,000 joules or 3.6 megajoules; The kinetic energy of a 2 kg mass moving at a velocity of 1 m/s. The energy is linear in the mass but quadratic in the velocity, being given by $E = ½mv^2$;

Kinetic Energy

The **kinetic energy** of an object is the extra energy which it possesses due to its motion. It is defined as *the work needed to accelerate a body of a given mass from rest to its current velocity*. Having gained this energy during its acceleration, the body maintains this kinetic energy unless its speed changes. Negative work of the same magnitude would be required to return the body to a state of rest from that velocity.

Laminar Jets

Laminar flow, sometimes known as streamline flow, occurs when a fluid flows in parallel layers, with no disruption between the layers. In fluid dynamics, laminar flow is a flow regime characterized by high momentum diffusion, low momentum convection, pressure and velocity independent from time. It is the opposite of turbulent flow. In nonscientific terms laminar flow is "smooth," while turbulent flow is "rough."

Law of Conservation of Energy

In physics, the **law of conservation of energy** states that the total amount of energy in an isolated system remains constant. A consequence of this law is that energy cannot be created or destroyed. The only thing that can happen with energy in an isolated system is that it can change form, that is to say for instance kinetic energy can become thermal energy. Another consequence of this law is that perpetual motion machines can only work if they deliver no energy to their surroundings, and that devices that produce more energy than is put into them are **impossible.**

Laws of Thermodynamics

The **laws of thermodynamics**, in principle, describe the specifics for the transport of heat and work in thermodynamic processes. Since their inception, however, these laws have become some of the most important in all of physics and other branches of science connected to thermodynamics.

There are generally considered to be four laws:

The zeroth law of thermodynamics, which underlies the definition of temperature.

The **first law** of thermodynamics, which mandates conservation of energy, and states in particular that heat is a form of energy.

The **second law** of thermodynamics, which states that the entropy of the universe always increases, or (equivalently) that perpetual motion machines are **impossible.**

The **third law** of thermodynamics, which concerns the entropy of an object at absolute zero temperature, and implies that it is impossible to cool a system all the way to exactly absolute zero.

Over the last 80 years or so, occasionally, various writers have suggested adding Laws, but none of them have been widely accepted.

Lithium-Ion Batteries

Lithium-ion batteries (sometimes abbreviated **Li-ion batteries**) are a type of rechargeable battery in which a lithium ion moves between the anode and cathode. The lithium ion moves from the anode to the cathode during discharge and from the cathode to the anode when charging. Lithium ion batteries are commonly used in consumer electronics. They are currently one of the most popular types of battery for portable electronics, with one of the best energy-to-weight ratios, no memory effect, and a slow loss of charge when not in use. In addition to uses for consumer electronics, lithium-ion batteries are growing in popularity for defense, automotive, and aerospace applications due to their high energy density. However certain kinds of mistreatment may cause Li-ion batteries to explode. The three primary functional components of a lithium ion battery are the anode, cathode, and electrolyte, for which a variety of materials may be used. Commercially, the most popular material for the anode is graphite. The cathode is generally one of three materials: a layered oxide, such as lithium cobalt oxide, one based on a polyanion, such as lithium iron phosphate, or a spinel, such as lithium manganese oxide, although materials such as TiS_2 (titanium disulfide) were originally used.[3] Depending on the choice of material for the anode, cathode, and electrolyte the voltage, capacity, life, and safety of a lithium ion battery can change dramatically. Lithium ion batteries are not to be confused with lithium batteries, the key difference being that lithium batteries are primary batteries containing metallic lithium while lithium-ion batteries are secondary batteries containing an intercalation anode material.

Local Seething Vacuum

Every electrical circuit ever built — and those built today — are in fact powered by electrical energy extracted by the circuit dipolarity from the local seething vacuum, from active space itself. But our engineers are trained to build circuits which also self-destroy the extraction of that vacuum energy, faster than they power their loads.

Lorentz's Symmetrization of the Heavyside Equations in 1892

Lorentz–Heaviside units (or Heaviside–Lorentz units) for Maxwell's equations are often used in relativistic calculations. They differ from the equations in CGS units by a factor of $\sqrt{4\pi}$ in the definitions of the fields and electric charge. The units are particularly convenient when performing calculations in spatial dimensions greater than three such as is done in string theory.

Maxwell's equations with sources

The equations with sources take the following form:

$$\nabla \cdot \mathbf{E} = \rho$$
$$\nabla \cdot \mathbf{B} = 0$$

$$\nabla \times \mathbf{E} = -\frac{1}{c}\frac{\partial \mathbf{B}}{\partial t}$$
$$\nabla \times \mathbf{B} = \frac{1}{c}\frac{\partial \mathbf{E}}{\partial t} + \frac{1}{c}\mathbf{J}$$

where c is the speed of light in a vacuum. Here **E** is the electric field, **B** is the magnetic field, ϱ is the charge density, and **J** is the current density.

The charge and fields in Lorentz–Heaviside units are related to the quantities in cgs units by

$$q_{LH} \stackrel{\text{def}}{=} \sqrt{4\pi} q_{cgs}$$
$$\mathbf{E}_{LH} \stackrel{\text{def}}{=} \frac{\mathbf{E}_{cgs}}{\sqrt{4\pi}}$$
$$\mathbf{B}_{LH} \stackrel{\text{def}}{=} \frac{\mathbf{B}_{cgs}}{\sqrt{4\pi}}.$$

Lorentz force

The force exerted upon a charged particle by the electric field and magnetic field is given in both cgs and Lorentz–Heaviside units by the Lorentz force equation:

$$\mathbf{F} = q(\mathbf{E} + \frac{\mathbf{v}}{c} \times \mathbf{B}),$$

where q is the charge on the particle and **v** is the particle velocity. The magnetic field **B** has the same units as the electric field **E**. In physics, the **Lorentz force** is the force on a point charge due to electromagnetic fields. It is given by the following equation in terms of the electric and magnetic fields:

$$\mathbf{F} = q[\mathbf{E} + (\mathbf{v} \times \mathbf{B})],$$

where
F is the force (in newtons)
E is the electric field (in volts per meter)
B is the magnetic field (in teslas)
q is the electric charge of the particle (in coulombs)
v is the instantaneous velocity of the particle (in meters per second)
× is the vector cross product

∇ and $\nabla\times$ are gradient and curl, respectively
or equivalently the following equation in terms of the vector potential and scalar potential:

$$\mathbf{F} = q(-\nabla\phi - \frac{\partial \mathbf{A}}{\partial t} + \mathbf{v} \times (\nabla \times \mathbf{A})),$$

where:
A and ϕ are the magnetic vector potential and electrostatic potential, respectively, which are related to **E** and **B** by

$$\mathbf{E} = -\nabla\phi - \frac{\partial \mathbf{A}}{\partial t}$$

$$\mathbf{B} = \nabla \times \mathbf{A}.$$

Note that these are vector equations: All the quantities written in boldface are vectors (in particular, **F**, **E**, **v**, **B**, **A**). The Lorentz force law has a close relationship with Faraday's law of induction. A positively charged particle will be accelerated in the *same* linear orientation as the **E** field, but will curve perpendicularly to both the instantaneous velocity vector **v** and the **B** field according to the right-hand rule (in detail, if the thumb of the right hand points along **v** and the index finger along **B**, then the middle finger points along **F**). The term $q\mathbf{E}$ is called the **electric force**, while the term $q\mathbf{v} \times \mathbf{B}$ is called the **magnetic force**. According to some definitions, the term "Lorentz force" refers specifically to the formula for the magnetic force.

$$\mathbf{F}_{mag} = q(\mathbf{v} \times \mathbf{B})$$

with the *total* electromagnetic force (including the electric force) given some other (nonstandard) name. This article will *not* follow this nomenclature: In what follows, the term "Lorentz force" will refer only to the expression for the total force. The magnetic force component of the Lorentz force manifests itself as the force that acts on a current-carrying wire in a magnetic field. In that context, it is also called the **Laplace force**.

Magnetic Air Cars

Briefly, a magnetic air car embodiment of the present invention uses a magnetic motor to compress input air and save moderately compressed high-pressure air in storage tanks. The compressor and storage tanks deliver the high-pressure working air and operational flows to several stages of compressors that boost the pressures during driving to very high-pressure, then ultra high-pressure, then super high-pressure, and finally to extremely high-pressure. A pneumatic torque converter uses jets of the extremely high-pressure to turn an input shaft of a transmission and differential. These, in turn, drive the powered wheels of a car. The compressors float a connecting shaft with matching vanes and impellers on opposite ends on air bearings to reduce shaft turning friction to near zero. The balance of forces between the two ends of a coupled turbo pair allows a simple air bearing design to operate safely and reliably at high rotational speeds. An alternative automobile embodiment of the present invention comprises a magnetic engine that drives an air compressor to fill on-board compressed air tanks. The magnetic engine is started initially by an electric starter motor and battery. Air bearings are used throughout to reduce friction to near zero in the compressors, compressor, magnetic engine, and air cycle engine. The compressed air tanks are filled to only moderate pressures, such as 200-PSI. Compressed air from the tanks and the air compressor is passed through air amplifiers to greatly increase the volume of input air. Amplified air is input to two compressors that are each ducted and driven to function as two-stage compressors. High-pressure jets from these compressors are applied to a pneumatic engine that converts it to rotating mechanical energy to drive the wheels of a car. Exhaust heat and compressed air are recycled by a recuperator to improve efficiency. These and other objects and advantages of the present invention will no doubt become obvious to those of ordinary skill in the art after having read the following detailed description of the preferred embodiments which are illustrated in the drawing figures.

Magnetic Dipoles

The electromagnetic phenomenon. For the radio antenna, see *dipole antenna*. For the permanent magnet, see *dipole magnet*. For the graph, see *dipole graph*. For the loudspeaker enclosure, see *dipole speaker*.

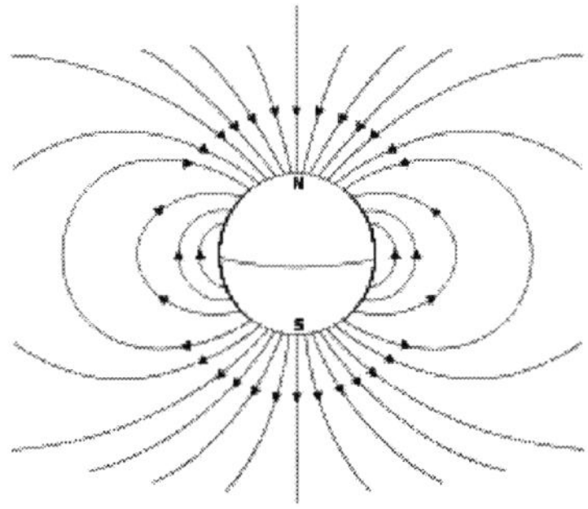

The Earth's magnetic field, which is approximately a magnetic dipole. However, the "N" and "S" (north and south) poles are labeled here geographically, which is the opposite of the convention for labeling the poles of a magnetic dipole moment In physics, there are two kinds of **dipoles** (Hellènic: di(s)- = two- and pòla = pivot, hinge): An **electric dipole** is a separation of positive and negative charge. The simplest example of this is a pair of electric charges of equal magnitude but opposite sign, separated by some, usually small, distance. A permanent electric dipole is called an electret. A **magnetic dipole** is a closed circulation of electric current. A simple example of this is a single loop of wire with some constant current flowing through it.[1][2]

Dipoles can be characterized by their dipole moment, a vector quantity. For the simple electric dipole given above, the electric dipole moment would point from the negative charge towards the positive charge, and have a magnitude equal to the strength of each charge times the separation between the charges. For the current loop, the magnetic dipole moment would point through the loop (according to the right hand grip rule), with a magnitude equal to the current in the loop times the area of the loop. In addition to current loops, the electron, among other fundamental particles, is said to have a magnetic dipole moment. This is because it generates a magnetic field which is identical to that generated by a very small current loop. However, to the best of our knowledge, the electron's magnetic moment is not due to a current loop, but is instead an intrinsic property of the electron. It is also possible that the electron has an electric dipole moment, although this has not yet been observed (see electron electric dipole moment for more information).

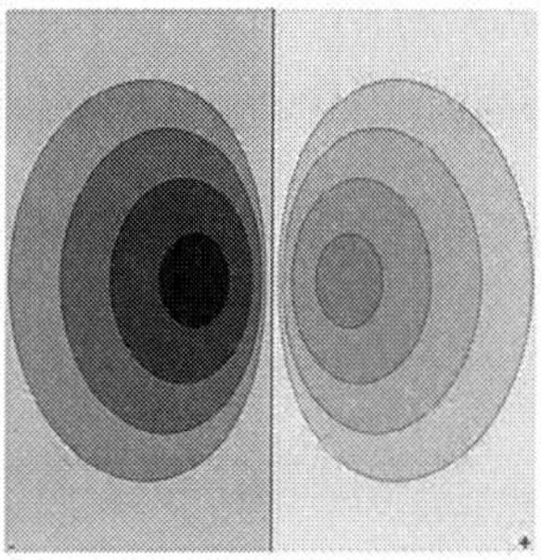

Contour plot of an electrical dipole, with equipotential surfaces indicated A permanent magnet, such as a bar magnet, owes its magnetism to the intrinsic magnetic dipole moment of the electron. The two ends of a bar magnet are referred to as poles (not to be confused with monopoles), and are labeled "north" and "south." The dipole moment of the bar magnet points from its magnetic south to its magnetic north pole—confusingly, the "north" and "south" convention for magnetic dipoles is the opposite of that used to describe the Earth's geographic and magnetic poles, so that the Earth's geomagnetic north pole is the *south* pole of its dipole moment. (This should not be difficult to remember; it simply means that the north pole of a bar magnet is the one which points north if used as a compass.) The only known mechanisms for the creation of magnetic dipoles are by current loops or quantum-mechanical spin since the existence of magnetic monopoles has never been experimentally demonstrated.

Magnetic Field

In physics, a *magnetic field* is a vector field that permeates space and which can exert a *magnetic force* on moving electric charges and on magnetic dipoles (such as permanent magnets). When placed in a magnetic field, magnetic dipoles tend to align their axes to be parallel with the magnetic field, as can be seen when iron filings are in the presence of a magnet (see picture at right). In addition, a changing magnetic field can induce an electric field. Magnetic fields surround and are created by electric currents, magnetic dipoles, and changing electric fields. Magnetic fields also have their own energy, with an energy density proportional to the square of the field intensity. There are some notable specific instances of the magnetic field. For the physics of magnetic materials, see magnetism and magnet, and more specifically ferromagnetism, paramagnetism, and diamagnetism. For constant magnetic fields, such as are generated by stationary dipoles and steady currents, see magnetostatics. For magnetic fields created by changing electric fields, see electromagnetism. The electric field and the magnetic field are tightly interlinked, in two senses. First, changes in either of these fields will cause ("induce") changes in the other, according to Maxwell's equations. Second, according to Einstein's theory of special relativity, a magnetic force in one inertial frame of reference may be an electric force in another, or vice-versa (see relativistic electromagnetism for examples). Together, these two fields make up the electromagnetic field, which is best known for underlying light and other electromagnetic waves.

Magnetic Flux

Magnetic flux, represented by the Greek letter Φ (phi), is a measure of quantity of magnetism, taking into account the strength and the extent of a magnetic field. The SI unit of magnetic flux is the weber (in derived units: volt-seconds), and the unit of magnetic field is the weber per square meter, or tesla.

Magnetic Motors

A **magnetic motor** (or **magnet motor**) is a device which converts power of or relating to or caused by magnetism (e.g., "magnetic forces") into mechanical force and motion, with no other input. It usually provides rotary mechanical motion. The machines that utilizes the properties of a magnet for mechanical energy.

Matter in Motion

While in the popular mind, **eternity** often simply means existing for a limitless amount of time, many have used it to refer to a timeless existence altogether outside of time. There are a number of arguments for eternity, by which proponents of the concept, principally Aristotle, purported to prove that matter, motion, and time must have existed eternally.

Maxwell's Equations

In electromagnetism, **Maxwell's equations** are a set of four partial differential equations that describe the properties of the electric and magnetic fields and relate them to their sources, charge density and current density. These equations are used to show that light is an electromagnetic wave. Individually, the equations are known as Gauss's law, Gauss's law for magnetism, Faraday's law of induction, and Ampère's law with Maxwell's correction.

These four equations, together with the Lorentz force law are the complete set of laws of classical electromagnetism. The Lorentz force law itself was actually derived by Maxwell under the name of "Equation for Electromotive Force" and was one of an earlier set of eight Maxwell's equations.

Mechanical Energy

Scientists make simplifying assumptions to make calculations about how mechanical systems react. For example, instead of calculating the mechanical energy separately for each of the billions of molecules in a soccer ball, it is easier to treat the entire ball as one object. This means that only two numbers (one for kinetic mechanical energy, and one for potential mechanical energy) are needed for each dimension (for example, up/down, north/south, east/west) under consideration. To calculate the energy of a system without any simplifying assumptions would require examining the state of all elementary particle(s) and considering all four fundamental interactions). This is usually only done for very small systems, such as those studied in particle physics.

Milankovitch-Like Wobbles

Milankovitch cycles are the collective effect of changes in the Earth's movements upon its climate, named after Serbian civil engineer and mathematician Milutin Milanković. The eccentricity, axial tilt, and precession of the Earth's orbit vary in several patterns, resulting in 100,000-year ice age cycles of the Quaternary glaciation over the last few million years. The Earth's axis completes one full cycle of precession approximately every 26,000 years. At the same time, the elliptical orbit rotates, more slowly, leading to a 21,000-year cycle between the seasons and the orbit. In addition, the angle between Earth's rotational axis and the normal to the plane of its orbit moves from 22.1 degrees to 24.5 degrees and back again on a 41,000-year cycle. Currently, this angle is 23.44 degrees and is decreasing. The Milankovitch theory of climate change is not perfectly worked out; in particular, the largest observed response is at the 100,000-year timescale, but the forcing is apparently small at this scale, in regard to the ice ages. Various feedbacks (from carbon dioxide, or from ice sheet dynamics) are invoked to explain this discrepancy. Milankovitch-like theories were advanced by Joseph Adhemar, James Croll and others, but verification was difficult due to the absence of reliably dated evidence and doubts as to exactly which periods were important. Not until the advent of deep-ocean cores and a seminal paper by Hays, Imbrie and Shackleton, "Variations in the Earth's Orbit: Pacemaker of the Ice Ages", in *Science*, 1976, did the theory attain its present state.

Nanocrystalline

Nanocrystalline silicon (nc-Si), sometimes also known as **microcrystalline silicon (μc-Si)**, is a form of porous silicon [1]. It is an allotropic form of silicon with paracrystalline structure—is similar to amorphous silicon (a-Si), in that it has an amorphous phase. Where they differ, however, is that nc-Si has small grains of crystalline silicon within the amorphous phase. This is in contrast to polycrystalline silicon (poly-Si) which consists solely of crystalline silicon grains, separated by grain boundaries. . The difference comes solely from the grain size of the crystalline grains. Most materials with grains in the micrometre range are actually fine-grained polysilicon, so nanocrystalline silicon is a better term. The term Nanocrystalline silicon refers to a range of materials around the transition region from amorphous to microcrystalline phase in the silicon thin film. The crystalline volume fraction (as measured from Raman spectroscopy) is another criteriòn to describe the materials in this transition zone. nc-Si has many useful advantages over a-Si, one being that if grown properly it can have a higher electron mobility, due to the presence of the silicon crystallites. It also shows increased absorption in the red and infrared wavelengths, which make it an important material for use in a-Si solar cells. One of the most important advantages of nanocrystalline silicon, however, is that it has increased stability over a-Si, one of the reasons being because of its lower hydrogen concentration.[citation needed] Although it currently cannot attain the mobility that poly-Si can, it has the advantage over poly-Si that it is easier to fabricate, as it can be deposited using conventional low temperature a-Si deposition techniques, such as PECVD, as opposed to laser annealing or high temperature CVD processes, in the case of poly-Si.

NdFEB

A **neodymium magnet** or **NIB magnet** (a variety of rare-earth magnet) is a powerful magnet made of a combination of neodymium, iron, and boron — $Nd_2Fe_{14}B$.

Negative Charge/ Positive Charge

Electric charge is a fundamental conserved property of some subatomic particles, which determines their electromagnetic interaction. Electrically charged matter is influenced by, and produces, electromagnetic fields. The interaction between a moving charge and an electromagnetic field is the source of the electromagnetic force, which is one of the four fundamental forces. Electric charge is a characteristic of some subatomic particles. It is quantized in that, when expressed in units of the so-called elementary charge *e*, it takes integer or fractional values. Electrons by convention have a charge of −1, while protons have the opposite charge of +1. Quarks have a fractional charge of $-\frac{1}{3}$ or $+\frac{2}{3}$. The antiparticle equivalents of these (positrons, antiprotons, and antiquarks, respectively) have the opposite charge. There are other charged particles. The discrete nature of electric charge was proposed by Michael Faraday in his electrolysis experiments, and then directly demonstrated by Robert Millikan in his oil-drop experiment. In general, same-sign charged particles repel one another, while different-sign charged particles attract. This is expressed quantitatively in Coulomb's law, which states that the magnitude of the electrostatic repelling force between two particles is proportional to the product of their charges and the inverse square of the distance between them. The electric charge of a macroscopic object is the sum of the electric charges of its constituent particles. Often, the net electric charge is zero, because it is favorable for the number of electrons in every atom to equal the number of protons (or, more generally, for the number of anions, or negatively charged atoms, in every molecule to equal the number of cations, or positively charged atoms). When the net electric charge is non-zero and motionless, one has the phenomenon known as static electricity. Even when the net charge is zero, it can be distributed non-uniformly (e.g., due to an external electric field, or due to molecular motion), in which case the material is said to be polarized. The charge due to the polarization is known as bound charge, while the excess charge brought from outside is called *free charge*. The motion of charged particles (e.g., of electrons in metals) in a particular direction is known as electric current. The SI unit of quantity of electric charge is the coulomb, which is equivalent to about 6.25×10^{18} e (the charge on a single electron or proton). Hence, the charge of an electron is approximately -1.602×10^{-19} C. The coulomb is defined as the quantity of charge that has passed through the cross-section of an electrical conductor carrying one ampere within one second. The symbol *Q* is often used to denote a quantity of electricity or charge. The quantity of electric charge can be directly measured with an electrometer, or indirectly measured with a ballistic galvanometer. After finding the quantized character of charge, in 1891 Stoney proposed the unit 'electron' for this fundamental unit of electrical charge. This was before the discovery of the particle by J.J. Thomson in 1897. Today, the name "electron" for the unit of charge is no longer widely used except in the derived unit "electronvolt". This is quite surprising considering the wide use of this unit in the fields of physics and chemistry. The unit is today treated as nameless, referred to as "fundamental unit of charge" or simply as "e". Formally, a measure of charge should be a multiple of the elementary charge *e* (charge is quantized), but since it is an average, macroscopic quantity, many orders of magnitude larger than a single elementary charge, it can effectively take on any real value. Furthermore, in some contexts it is meaningful to speak of fractions of a charge; e.g. in the charging of a capacitor.

Neodymium Magnets

Neodymium magnets are very strong relative to their mass, but are also mechanically fragile. Like other ferromagnetic materials, neodymium magnets lose their magnetism above a temperature known as the Curie point. The most powerful grades have a maximum working temperature at a relatively low temperature: below 80 degrees Celsius (176 degrees Fahrenheit). High-temperature grades will operate at up to 200 and even 230°C but their strength is only marginally greater than that of a samarium-cobalt magnet. As of 2008 neodymium magnets cost about $44/kg, $1.40 per BHmax. Most neodymium magnets are anisotropic, and hence can only be magnetised along one direction although B10N material is isotropic. During manufacture fields of 30-40 kOe are required to saturate the material. Neodymium magnets have a coercivity (required demagnetisation field from saturation) of about 10,000-12,000 Oersted. **Range of strength**: Neodymium magnets (or "neo" as they are known in the industry) are graded in strength from N24 to the strongest, N55. The theoretical limit for neodymium magnets is grade N64. The number after the N represents the magnetic energy product, in megagauss-oersteds (MGOe) (1 MG·Oe = $7{,}958 \cdot 10^3$ T·A/m = 7,958 kJ/m^3). N48 has a remnant static magnetic field of 1.38 teslas and an *H* (magnetic field intensity) of 13,800 Oersteds (1.098 MA/m). By volume one requires about 18 times as much ceramic magnetic material for the equivalent magnet lifting strength, and about 3 to 5 times as much for the equivalent dipole moment. A neodymium magnet holding up 1300 times its own weight. **Uses:** They have replaced marginally weaker and significantly more heat-resistant samarium-cobalt magnets in most applications, due mainly to their lower cost. Used for stabilization and angular head motors in computer hard drives, neodymium magnets are also popular with hobbyists, and a small magnet can have amazing properties — it exhibits magnetic braking when moved near a non-magnetic metal due to induced eddy currents. An excellent demonstration for students to see the effects of Lenz's Law in non-ferrous metals may be performed by dropping a strong neodymium magnet through a copper pipe. The magnet will travel through the pipe remarkably slowly as it falls. The effect may be greatly enhanced by immersing the pipe in liquid nitrogen (thus increasing its conductivity even further) prior to dropping the magnet through. A somewhat larger magnet interacts strongly enough with the magnetic field of the Earth to allow its tendency to align with that field to be perceived directly when holding it, essentially forming a compass. Cylinder- and disc-shaped neodymium magnets are especially responsive to the Earth's magnetic fields. Neodymium magnets are used for the transducers in many headphones. Neodymium magnets are becoming increasingly common in loudspeakers for high-volume sound reinforcement applications. Neodymium magnets have been used several times before on the Discovery Channel television show *Mythbusters* in myth tests using magnetism where conventional magnets were not powerful enough. A series of 13 very large magnets of this type were proven to be able to alter the trajectory path of a steel bullet.

Over-Unity Machines

The conservation laws are particularly robust. Noether's theorem states that any conservation law can be derived from a corresponding continuous symmetry.[8] In other words, so long as the laws of physics (not simply the current understanding of them, but the actual laws, which may still be undiscovered) and the various physical constants remain invariant over time — so long as the laws of the universe are fixed — then the conservation laws *must* be true, in the sense that they follow from the presupposition using mathematical logic. To put it the other way around: if perpetual motion or "**overunity**" machines were possible, then most of what we believe to be true about physics, mathematics, or both would have to be false. This problem is exacerbated by the fact that mathematics is believed to be absolute, since its veracity is not dependent on anything that happens in the real world.

Particles

Particle physics is a branch of physics that studies the elementary constituents of matter and radiation, and the interactions between them. It is also called **high energy physics**, because many elementary particles do not occur under normal circumstances in nature, but can be created and detected during energetic collisions of other particles, as is done in particle accelerators. Research in this area has produced a long list of particles. Modern particle physics research is focused on subatomic particles, which have less structure than atoms. These include atomic constituents such as electrons, protons, and neutrons (protons and neutrons are actually composite particles, made up of quarks), particles produced by radiative and scattering processes, such as photons, neutrinos, and muons, as well as a wide range of exotic particles.

Strictly speaking, the term *particle* is a misnomer because the dynamics of particle physics are governed by quantum mechanics. As such, they exhibit wave-particle duality, displaying particle-like behavior under certain experimental conditions and wave-like behavior in others (more technically they are described by state vectors in a Hilbert space; see quantum field theory). Following the convention of particle physicists, "elementary particles" refer to objects such as electrons and photons, with the understanding that these "particles" display wave-like properties as well. All the particles and their interactions observed to date can almost be described entirely by a quantum field theory called the Standard Model. The Standard Model has 40 species of elementary particles (24 fermions, 12 vector bosons, and 4 scalar bosons), which can combine to form composite particles, accounting for the hundreds of other species of particles discovered since the 1960s. The Standard Model has been found to agree with almost all the experimental tests conducted to date. However, most particle physicists believe that it is an incomplete description of nature, and that a more fundamental theory awaits discovery. In recent years, measurements of neutrino mass have provided the first experimental deviations from the Standard Model. Particle physics has had a large impact on the philosophy of science. Some particle physicists adhere to reductionism, a point of view that has been criticized and defended by philosophers and scientists. Part of the debate is described below.

History: The idea that all matter is composed of elementary particles dates to at least the 6th century BC. The philosophical doctrine of atomism and the nature of elementary particles were studied by ancient Greek philosophers such as Leucippus, Democritus and Epicurus; ancient Indian philosophers such as Kanada, Dignāga and Dharmakirti; medieval scientists such as Alhazen, Avicenna and Algazel; and early modern European physicists such as Pierre Gassendi, Robert Boyle and Isaac Newton. The particle theory of light was also proposed by Alhazen, Avicenna, Gassendi and Newton. These early ideas were founded in abstract, philosophical reasoning rather than experimentation and empirical observation. In the 19th century, John Dalton, through his work on stoichiometry, concluded that each element of nature was composed of a single, unique type of particle. Dalton and his contemporaries believed these were the fundamental particles of nature and thus named them atoms, after the Greek word *atomos*, meaning "indivisible". However, near the end of the century, physicists discovered that atoms were not, in fact, the fundamental particles of nature, but conglomerates of even smaller particles. The early 20th century explorations of nuclear physics and quantum physics culminated in proofs of nuclear fission in 1939 by Lise Meitner (based on experiments by Otto Hahn), and nuclear fusion by Hans Bethe in the same year. These discoveries gave rise to an active industry of generating one atom from another, even rendering possible (although not profitable) the transmutation of lead into gold. They also led to the development of nuclear weapons. Throughout the 1950s and 1960s, a bewildering variety of particles were found in scattering experiments. This was referred to as the "particle zoo". This term was deprecated after the formulation of the Standard Model during the 1970s in which the large number of particles was explained as combinations of a (relatively) small number of fundamental particles.

Pelton-Type Impulse turbine

The Pelton wheel is among the most efficient types of water turbines. It was invented by Lester Allan Pelton (1829-1908) in the 1870s, and is an impulse machine, meaning that it uses the principle of Newton's second law to extract energy from a jet of fluid. Although the one-piece cast impulse water turbine was invented by Samuel Knight in Sutter Creek, in the California Mother Lode gold mining region, Pelton modified this invention to create his more efficient design. Knight Foundry is the last water-powered foundry known to exist in the United States and is still operated using Knight impulse turbines, used to extract power from high heads and low discharge water flows. The water flows along the tangent to the path of the runner. Nozzles direct forceful streams of water against a series of spoon-shaped buckets mounted around the edge of a wheel. As water flows into the bucket, the direction of the water velocity changes to follow the contour of the bucket. When the water-jet contacts the bucket, the water exerts pressure on the bucket and the water is decelerated as it does a "u-turn" and flows out the other side of the bucket at low velocity. In the process, the water's momentum is transferred to the turbine. This "impulse" does work on the turbine. For maximum power and efficiency, the turbine system is designed such that the water-jet velocity is twice the velocity of the bucket. A very small percentage of the water's original kinetic energy will still remain in the water; however, this allows the bucket to be emptied at the same rate it is filled, (see conservation of mass), thus allowing the water flow to continue uninterrupted. Often two buckets are mounted side-by-side, thus splitting the water jet in half (see photo). This balances the side-load forces on the wheel, and helps to ensure smooth, efficient momentum transfer of the fluid jet to the turbine wheel. Because water and most liquids are nearly incompressible, almost all of the available energy is extracted in the first stage of the hydraulic turbine. Therefore, Pelton wheels have only one turbine stage, unlike gas turbines that operate with compressible fluid. Applications

Pelton wheels are the preferred turbine for hydro-power, when the available water source has relatively high hydraulic head at low flow rates. Pelton wheels are made in all sizes. There exist multi-ton Pelton wheels mounted on vertical oil pad bearings in hydroelectric plants. The largest units can be up to 200 megawatts. The smallest Pelton wheels are only a few inches across, and can be used to tap power from mountain streams having flows of a few gallons per minute. Some of these systems utilize household plumbing fixtures for water delivery. These small units are recommended for use with thirty metres or more of head, in order to generate significant power levels. Depending on water flow and design, Pelton wheels operate best with heads from 15 metres to 1,800 metres, although there is no theoretical limit. The Pelton wheel is most efficient in high head applications. Thus, more power can be extracted from a water source with high-pressure and low-flow than from a source with low-pressure and high-flow, even though the two flows theoretically contain the same power. Also a comparable amount of pipe material is required for each of the two sources, one requiring a long thin pipe, and the other a short wide pipe.

Permanent Magnets

Magnet (from Greek μαγνήτης λίθος, "Magnesian stone") is a material or object that produces a magnetic field. This magnetic field is invisible but is responsible for the most notable property of a magnet: a force that pulls on other ferromagnetic materials and attracts or repels other magnets. A **permanent magnet** is one that stays magnetized, such as a magnet used to hold notes on a refrigerator door. Materials which can be magnetized, which are also the ones that are strongly attracted to a magnet, are called ferromagnetic. These include iron, nickel, cobalt, some rare earth metals and some of their alloys, and some naturally occurring minerals such as lodestone. Permanent magnets are made from "hard" ferromagnetic materials which are designed to stay magnetized, while "soft" ferromagnetic materials like soft iron are attracted to a magnet but don't tend to stay magnetized. An **electromagnet** is made from a coil of wire which acts as a magnet when an electric current passes through it, but stops being a magnet when the current stops. Often an electromagnet is wrapped around a core of ferromagnetic material like steel, which enhances the magnetic field produced by the coil. Although ferromagnetic materials are the only ones strongly enough attracted to a magnet to be commonly considered "magnetic", all other substances respond weakly to a magnetic field, by one of several other types of magnetism. Paramagnetic materials, such as aluminum and oxygen are weakly attracted to a magnet. Diamagnetic materials, such as carbon and water, which include all substances not having another type of magnetism, are weakly repelled by a magnet. The overall strength of a magnet is measured by its magnetic moment, while the local strength of the magnetism in a material is measured by its magnetization.

Perpetual Motion Machines (PMM)

The term **perpetual motion**, taken literally, refers to movement that goes on forever. However, the term more generally refers to any closed system that produces more energy than it consumes. Such a device or system would be in violation of the law of conservation of energy, which states that energy can never be created or destroyed. The most conventional type of perpetual motion machine is a mechanical system which (supposedly) sustains motion despite losing energy to friction and air resistance, or while avoiding losing energy to friction and air resistance. According to the law of conservation of energy, such a device cannot exist.

Perpetual motion violates either the first law of thermodynamics, the second law of thermodynamics, or both. The first law of thermodynamics is essentially a statement of conservation of energy. The second law can be phrased in several different ways, the most intuitive of which is that heat flows spontaneously from hotter to colder places; the most well known statement is that entropy tends to increase, or at the least stay the same; another statement is that no heat engine (an engine which produces work while moving heat between two places) can be more efficient than a Carnot heat engine. As a special case of this, any machine operating in a closed cycle cannot *only* transform thermal energy to work in a region of constant temperature. Machines which are claimed not to violate either of the two laws of thermodynamics but rather are claimed to generate energy from unconventional sources are sometimes referred to as perpetual motion machines, although they are generally reported as not meeting the standard criteria for the name. By way of example, it is quite possible to design a clock or other low-power machine to run on the differences in barometric pressure or temperature between night and day. Such a machine has a source of energy, albeit one from which it is quite impractical to produce power in quantity. Classification: It is customary to classify perpetual motion machines according to which law of thermodynamics it attempts to violate: A **perpetual motion machine of the first kind** produces energy from nothing, giving the user unlimited 'free' energy. It thus violates the law of conservation of energy. A **perpetual motion machine of the second kind** is a machine which spontaneously converts thermal energy into mechanical work. When the thermal energy is equivalent to the work done, this does not violate the law of conservation of energy. However it does violate the more subtle second law of thermodynamics (see also entropy). Note that such a machine is different from real heat engines (such as car engines), which always involve a transfer of heat from a hotter reservoir to a colder one, the latter being warmed up in the process. The signature of a perpetual motion machine of the second kind is that there is only one single heat reservoir involved, which is being spontaneously cooled without involving a transfer of heat to a cooler reservoir. This conversion of heat into useful work, without any side effect, is impossible by the second law of thermodynamics. What may prove more useful is to explain the existence of hot reservoirs to begin with. A hot reservoir inside an internal combustion engine is created by a spark igniting fumes which contain stores of chemical energy. The temperature of the fumes increases above that of the surroundings. This is **not** a perpetual motion machine since the ability to raise the temperature above that of the surroundings depends on finite chemical reactions always less than the total heat energy and mass-energy contained within the system. Since there are far more states in which heat distribution is closer to thermodynamic equilibrium than states in which heat is concentrated in small regions, heat will tend to smooth out over time to lower power densities of increasingly unusable forms. A more obscure category is a **perpetual motion machine of the third kind**, usually (but not always) defined as one that completely eliminates friction and other dissipative forces, to maintain motion forever (due to its mass inertia). *Third* in this case refers solely to place in the above classification scheme, not the third law of thermodynamics. Although it is impossible to make such a machine, as dissipation can never be 100% eliminated in a mechanical system, it is nevertheless possible to get very close to this ideal (see examples in the Low Friction section). Unfortunately, even if such a machine could be built, it would not serve as an endless source of energy, since any energy-extracting mechanism would also serve as a dissipative force. If we could build a frictionless flywheel, it would eventually slow down & stop completely if its kinetic energy were tapped for useful work, and we would get no more energy out than the amount that was initially put in to spin up the flywheel. Use of the term "impossible" and perpetual motion Like all scientific theories, the laws of physics are incomplete. "A world that was simple enough to be fully known would be too simple to contain conscious observers that might know it." Outside of pure mathematics, stating that things are absolutely impossible is more a hallmark of pseudoscience than of true science. Nevertheless, the term is properly used to reflect those things that cannot be true without a significant rewrite of nearly all known scientific laws. The conservation laws are particularly robust. Noether's theorem states that any conservation law can be derived from a corresponding continuous symmetry. In other words, so long as the laws of physics (not simply the current understanding of them, but the actual laws, which may still be undiscovered) and the various physical constants remain invariant over time — so long as the laws of the universe are fixed — then the conservation laws *must* be true, in the sense that they follow from the presupposition using mathematical logic. To put it the other way around: if perpetual motion or "**overunity**" machines were possible, then most of what we believe to be true about physics, mathematics, or both would have to be false. This problem is exacerbated by the fact that mathematics is believed to be absolute, since its veracity is not dependent on anything that happens in the real world.

In this case it is easy to check whether or not the theory is correct. Using telescopes we can examine the universe in the distant past; the fact that stars even exist and are, to the limits of our measurements, identical to stars today, is a direct visual demonstration that physics was similar in the past. Combining different measurements such as spectroscopy, direct measurement of the speed of light in the past and similar measurements demonstrates conclusively that physics has remained substantially the same, if not identical, for all of observable history spanning billions of years. The principles of thermodynamics are so well established, both theoretically and experimentally, that proposals for perpetual motion machines are universally met with disbelief on the part of physicists. Any proposed perpetual motion design offers a potentially instructive challenge to physicists: one is almost completely certain that it can't work, so one must explain *how* it fails to work. The difficulty (and the value) of such an exercise depends on the subtlety of the proposal; the best ones tend to arise from physicists' own thought experiments and often shed light into unique aspects of physics. The law that entropy always increases, holds, I think, the supreme position among the laws of Nature. If someone points out to you that your pet theory of the universe is in disagreement with Maxwell's equations — then so much the worse for Maxwell's equations. If it is found to be contradicted by observation — well, these experimentalists do bungle things sometimes. But if your theory is found to be against the *second law of thermodynamics* I can give you no hope; there is nothing for it but to collapse in deepest humiliation. — Sir Arthur Stanley Eddington, *The Nature of the Physical World* (1927)

Photovoltaic Solar Cells

A **photovoltaic system** is a system which uses solar cells to convert light into electricity. A photovoltaic system consists of multiple components, including cells, mechanical and electrical connections and mountings and means of regulating and/or modifying the electrical output.

Due to the low voltage of an individual solar cell (typically ca. 0.5V), several cells are combined into photovoltaic modules, which are in turn connected together into an array. The electricity generated can be either stored, used directly (island/standalone plant)or fed into a large electricity grid powered by central generation plants (grid-connected/grid-tied plant) or combined with one or many domestic electricity generators to feed into a small grid (hybrid plant) [1]. Depending on the type of application, the rest of the system ("balance of system" or "BOS") consists of different components. The BOS depends on the load profile and the system type. Systems are generally designed in order to ensure the highest energy yield for a given investment.

PM Motors and Motor Generators

Better known in relatively small physical sizes and lower power applications, brushless permanent magnet (PM) motors can be made in virtually any size, with no real technological constraints. Large brushless PM motors are not particularly new. They're available from select manufacturers, which now seek to overcome past economic issues that have limited their numbers. Large brushless PM motors offer numerous benefits, among them high power density and high efficiency without rotor losses. Benefits come at a price, as manufacturing and material costs—including that of high-performance magnets—add up fast. Brushless PM motors also need a variable-frequency drive for control. Yet, cost-benefit figures favor PM motors over alternative technologies (see below) for a growing number of tough applications in industry, marine propulsion, military/defense, and other sectors. Thorough analysis of a potential application is essential to justify the cost-benefit. Recent developments may lend a helping hand. Widespread availability of drives, along with significant downward price trend for drives and magnet materials suggests more of these motors will be in service in the near future. Greater user awareness about large PM motors would further aid that development

Pneumatic Torque Converter

If you've read about manual transmissions, you know that an engine is connected to a transmission by way of a clutch. Without this connection, a car would not be able to come to a complete stop without killing the engine. But cars with an **automatic transmission** have no clutch that disconnects the transmission from the engine. Instead, they use an amazing device called a **torque converter**. It may not look like much, but there are some very interesting things going on inside. Torque converters may be electric or pneumatic

Polarized Vacuum

The most intense source of polarized vacuum UV and X radiation is synchrotron radiation, which exhibits a degree of partially polarized light between about 80-100%. However, the radiation transmitted by vacuum UV monochromators can also be highly polarized. The Seya-Namioka type of monochromator can produce partially polarized radiation between 50-80%. For certain experiments it is necessary to know the degree of polarization of the radiation being used. Also, when synchrotron radiation and a monochromator are combined the polarization characteristic of both should be known in order to make full use of these polarization properties. The polarizing effect of monochromators (i.e., diffraction gratings) have been measured at the Seya angle and at grazing angles for various spectral orders. Experimental evidence is presented which shows that the reciprocity law holds for polarization by reflection where the angle of incidence and diffraction are unequal. These results are reviewed along with the techniques for measuring the degree of polarization.

Pourous Media Technology

Air bearings are typically classed as 'orifice' or 'porous media' bearings. In orifice bearings, the pressurized air is supplied to the bearing surface through a small number of precisely-sized holes. The concept is similar to the air hockey table game, but with the holes in the puck rather than the table. Porous media air bearings are quite different in that the air is supplied through the entire surface of the bearing (Figure 6). The porous material controls the airflow in the same way an orifice bearing would do if it had millions of miniature holes across its surface. Broadly speaking, there are two techniques for achieving the compensating effect in air bearings. Orifice compensation is traditionally the most widely used method, but porous surface compensation is rapidly emerging as the method of choice due to its many advantages and increasing availability. In traditional orifice compensation, the precisely-sized orifices are strategically placed on the bearing and are often combined with grooves to distribute the pressurized air evenly across the bearing face. However, if the bearing face becomes scratched across a groove or near an orifice, the volume of air that escapes may be more than the orifice can supply, causing the bearing to crash even with normal air supply pressure. Rather than the small number of orifices of conventional air bearings, porous air bearings control the airflow across the entire bearing surface through millions of holes in the porous material. Because of this, they are harder to clog and will still fly even after being severely scratched. Even under normal operations, the air in an orifice air bearing loses pressure and creates pressure gradients in the air gap as it expands away from the orifice or groove. This is not an issue for porous air bearings where the air pressure remains almost uniform across the entire surface. Porous carbon has been found to be one of the best materials for this purpose, producing an ideal supply of uniform air pressure across the face of the bearing while automatically restricting and damping the air flow simultaneously. The carbon surface also provides greater bearing protection in the event of an air supply failure, and allows the bearings to be moved during air failure without damaging the support surface.

Poynting Energy Flow Theory S (S = E x H)

Professor Poynting, who first suggested that energy was distributed in space with a density E x H, also had a partial vision of the Heaviside signal. He definitely did not know that E is always perpendicular to H, and that the x in ExH means simply multiplication. (He had a term sinθ for the angle between them.) Poynting was writing before the general agreement that light is electromagnetic, and so he did not know that this Poynting energy always moved forward (in the third dimension) at a constant speed, 1/ [./(με)], the velocity of light in the medium. Poynting had a very good grasp of the direction of energy flow and its magnitude, but did not seem to understand the importance of reflections at a change of medium, which leads one to think of energy current ExH flowing backwards along the previous path, passing through the next portion of forwards travelling energy current. This superposition of forward and backward energy currents (implicit in the phrases "phase velocity" and "group velocity") has prevented a clear understanding of the electromagnetic wave. For fifty years, technology did not give us the power to drive the medium with an electromagnetic signal. With the low power at our disposal, all we could do was resonate the medium with periodic (sinusoidal) excitation in the same way as we move a child on a swing. In a resonant medium, energy is necessarily flowing in both directions; most of the forward energy returns to aid the source on the next cycle. Our inability to drive a medium except periodically insinuated itself into our group psyche, until we came to assert that nature *was* periodic (and even that it was sinusoidal). Implicit in this view were the wrong beliefs that

1. electromagnetic energy is necessarily contrapuntal,
2. E/H = ./(μ/ε) is not always true, (e.g. when two waves are passing through each other so that H cancels but E does not, so that E/H = infinity), and
3. Signals can travel slower than the speed of light 1/[./(με)] .

The absurdity of this third idea is easy to demonstrate if we consider a two directional highway. If all cars move at 60 m.p.h. but some (A per hour) move eastwards and some (B per hour) move westwards, no one would argue that the total passage of cars eastwards per hour pas a reference point, that is, (A-B), would help us to determine the velocity of cars by the formula

Flow of cars = (A-B) per hour
Distance between cars = L
Therefore velocity of cars = (A-B)L m.p.h.

However, this seems to be done, at least subconsciously, with phase velocity and group velocity. The very terms imply some such calculation. Some ten years ago the successful manufacture of high speed (1ns) logic elements capable of driving a 100 ohm load made it possible, for the first time for fifty years, to drive a medium rather than gently resonate it, as a matter of normal routine. Those driving a high speed logic step could clearly see it travelling at the speed of light for the dielectric (never any slower) and remaining unchanged on its journey. For the first time for seventy years, high speed digital engineers were privileged to see the Heaviside signal, an unchanging slab of ExH energy current guided between two conductors from one logic gate to the next. Reflections were prevented by proper termination at the destination, so that notions of phase velocity and group velocity evaporated. We saw a slab of energy launched from one pint, travelling unaltered, to be absorbed by the terminating resistor at the destination. At this point we just *had* to unburden ourselves at the theoretical level of implicit contrapuntal notions. A beautiful vision resulted, of a lateral strain ExH (where E/H = ./(μ/ε) which by definition travelled forward at velocity 1/ [./(με)].) As it travelled forward it filled (or probed) the space ahead of it in the same way as the ripples on the surface of a pond will fill the space (surface) as they come to it. Logic designers maintained a near constant aspect ratio in the space ahead, because whenever this slab came to a change in aspect ratio (= change of characteristic impedance, better termed characteristic resistance) some of the energy could double back on its tracks according to the well-known laws of reflection. However, this did not lead back to the old "phase velocity" and "group velocity" notions; rather, the slab of energy current split into two slabs, one to continue forward and the other to return, both slabs continuing to probe, or fill, the space presented to them on their journeys. The Heaviside signal offers us a dramatic simplification of our view of the fundamentals of electromagnetic theory.

Priority Valve

The priority valve remains closed until the engine changes pressure to vacuum. The vacuum lets the priority valve swing open. The engine now draws air as it needs. The priority valve inlet is a 3" tube. The end of the inlet is a 40 degree bologna cut. The valve chamber is a 3.5" square. The opening has a silicon seal. The valve door rests against the seal. When you have pressure on one side, it puts pressure against the seal and seals tighter. As the engine begins to draw, it changes pressure to vacuum. The priority valve doors swings open. The engine now draws air on its own. As a result, the star rotor is no longer needed to motivate the vehicle. Its discharges close. The pressure is now used to recharge the 2 ten-gallon tanks, and therefore, it revitalizes (recharges) itself. The throttle is achieved by governing the air pressure to air amplifier drives.

Rare Earth Elements (REE)

According to IUPAC, **rare earth elements** or **rare earth metals** are a collection of seventeen chemical elements in the periodic table, namely scandium, yttrium, and the fifteen lanthanoids. Scandium and yttrium are considered **rare earths** since they tend to occur in the same ore deposits as the lanthanoids and share similar chemical properties with them. Discovery and early history. Rare earth elements became known to the world with the discovery of the black mineral ytterbite (also known as gadolinite) by Lieutenant Carl Axel Arrhenius in the year 1787, in a quarry in the village of Ytterby, Sweden. Many of the rare earths are named in honor of the scientists who discovered or elucidated the elemental properties, geographical discovery, Latin or Greek, or mythology:

Name	Etymology
Lanthanum	from the Greek "lanthanon," meaning *I am hidden*.
Cerium	for the Roman deity of fertility Ceres.
Praseodymium	from the Greek "praso," meaning *leek-green*, and "didymos," meaning *twin*.
Neodymium	from the Greek "neo," meaning *new-one*, and "didymos," meaning *twin*.
Promethium	for the Titan Prometheus, who brought fire to mortals.
Samarium	for Vasili Samarsky-Bykhovets, who discovered the rare earth ore samarskite.
Europium	for the continent of Europe.
Gadolinium	for Johan Gadolin (1760-1852), to honor his investigation of rare earths.
Terbium	for the village of Ytterby, Sweden, where the first rare earth ore was discovered.
Dysprosium	from the Greek "dysprositos," meaning *hard to get*.
Holmium	for Stockholm (in Latin, "Holmia"), native city of one of its discoverers.
Erbium	for the village of Ytterby, Sweden.
Thulium	for the mythological land of Thule.
Ytterbium	for the village of Ytterby, Sweden.
Lutetium	for Lutetia, the city which later became Paris.

The term "Rare Earth" arises from the minerals from which they were isolated, which were uncommon oxide-type minerals (earths), only found in Gadolinite from one mine in the village of Ytterby, Sweden, However, with the exception of the highly-unstable promethium, the rare earth elements are found in relatively high concentrations in the earth's crust with Cerium being the 25th most abundant element in the earth's crust at 68 parts per million. The principal sources of rare earth elements are the minerals bastnäsite, monazite, and loparite and the lateritic ion-adsorption clays. Despite their high relative abundance, rare earth minerals are more difficult to mine and extract than equivalent sources of transition metals (due in part to their very similar chemical properties), making the rare earth elements relatively expensive. Their industrial use was very limited until efficient separation techniques were developed, such as ion exchange, fractional crystallization and liquid-liquid extraction during the late 50's and early 60's.

Abbreviations:

The following abbreviations are often used:
REE = rare earth elements
LREE = light rare earth elements (La-Sm)
HREE = heavy rare earth elements (Eu-Lu)

Technological applications:

Rare earth elements are incorporated into many modern technological devices, including superconductors, miniaturized magnets, electronic polishers, refining catalysts and hybrid car components. Rare earth ions are used as the active ions in luminescent materials used in optoelectronics applications, most notably the Nd:YAG laser. Phosphors with rare earth dopants are also widely used in cathode ray tube technology such as television sets. Global rare earth production: Up until 1948, most of the world's Rare Earths were sourced from placer sand deposits in India and Brazil. Through the 50's, South Africa then took the status as the world's Rare Earth source, after large Rare Earth bearing veins were discovered in Monazite. Today, those Indian and South African deposits still produce some Rare Earth concentrates, however they are dwarfed by the scale of Chinese production. China now produces over 95% of the world's Rare Earth supply. The use of rare earth elements in modern technology has increased dramatically over the past years. For example, dysprosium has gained significant importance for its use in the construction of hybrid car motors. Unfortunately, this new demand has strained supply, and there is growing concern that the world may soon face a shortage of the materials. All of the world's heavy rare earths (such as dysprosium) are sourced from Chinese Rare Earth sources such as the polymetallic Bayan Obo deposit.

High Rare Earth prices have wreaked havoc on many rural Chinese villages, as many illegal rare earth mines have been spewing toxic waste into the general water supply. Chinese export quotas have also resulted in a dramatic shift in the world's Rare Earth knowledge base. For example, the division of General Motors which deals with miniaturized magnet research recently shut down its US office and moved all of its staff to China. Geologic distribution: Due to lanthanide contraction, yttrium, which is trivalent, is of similar ionic size to dysprosium and its lanthanide neighbors. Due to the relatively gradual decrease in ionic size with increasing atomic number, the rare earth elements have always been difficult to separate. Even with eons of geological time, geochemical separation of the lanthanides has only rarely progressed much farther than a broad separation between light versus heavy lanthanides, otherwise known as the cerium and yttrium earths. This geochemical divide is reflected in the first two rare earths that were discovered, yttria in 1794 and ceria in 1803. As originally found, each comprised the entire mixture of the associated earths. Rare earth minerals, as found, usually are dominated by one group or the other, depending upon which size-range best fits the structural lattice. Thus, among the anhydrous rare earth phosphates, it is the tetragonal mineral xenotime that incorporates yttrium and the yttrium earths, whereas the monoclinic monazite phase incorporates cerium and the cerium earths preferentially. The smaller size of the yttrium group allows it a greater solid solubility in the rock-forming minerals that comprise the earth's mantle, and thus yttrium and the yttrium earths show less enrichment in the earth's crust, relative to chondritic abundance, than does cerium and the cerium earths. This has economic consequences: large orebodies of the cerium earths are known around the world, and are being actively exploited. Corresponding orebodies for yttrium tend to be rarer, smaller, and less concentrated. Most of the current supply of yttrium originates in the "ion adsorption clay" ores of Southern China. Some versions of these provide concentrates containing about 65% yttrium oxide, with the heavy lanthanides being present in ratios reflecting the Oddo-Harkins rule: even-numbered heavy lanthanides at abundances of about 5% each, and odd-numbered lanthanides at abundances of about 1% each. Similar compositions are found in xenotime or gadolinite.

Well-known minerals that contain yttrium include gadolinite, xenotime, samarskite, euxenite, fergusonite, yttrotantalite, yttrotungstite, yttrofluorite (a variety of fluorite), thalenite, yttrialite. Small amounts occur in zircon, which derives its typical yellow fluorescence from some of the accompanying heavy lanthanides. The zirconium mineral eudialyte, such as is found in southern Greenland, also contains small but potentially useful amounts of yttrium. Of the above yttrium minerals, most played a part in providing research quantities of lanthanides during the discovery days. Xenotime is occasionally recovered as a byproduct of heavy sand processing, but has never been nearly as abundant as the similarly recovered monazite (which typically contains a few percent of yttrium). Uranium ores processed in Ontario have occasionally yielded yttrium as a byproduct. Well-known minerals that contain cerium and the light lanthanoids include bastnaesite, monazite, allanite, loparite, ancylite, parisite, lanthanite, chevkinite, cerite, stillwellite, britholite, fluocerite, and cerianite. Over the years, monazite (marine sands from Brazil, India, or Australia; rock from South Africa), bastnaesite (from Mountain Pass California, or several localities in China), and loparite (Kola Peninsula, Russia) have been the principal ores of cerium and the light lanthanoids. A few sites are under development outside of China, the most significant of which are the Nolans Project in Central Australia, the remote Hoidas Lake project in northern Canada and the Mt. Weld project in Australia.[11] The Hoidas Lake project has the potential to supply about 10% of the $1 billion of REE consumption that occurs in North America every year.

Rear Wheel Pulley Systems

We want the alternator working with the air bearing pump/compressor when it is being utilized. On the final drive, we want a secondary alternator. They will both maintain battery output, and it can be any place. The second pulley system may not be necessary. The primary alternator will spin with the air bearing pump/compressor. The magnetic motor will spin both the air bearing pump compressor and the primary alternator. The secondary alternator can be spun from the final drive or have its own air drive. The alternator can have its own air bearings. Please understand that the alternator on the air bearing pump/compressor must be standard, unless we can provide an external air pressure source. Air bearings may be added for creating a zero-friction pulley, a common application for **creating zero-friction pulleys** for counter balances or Web handling.

Recapturing Braking Energy

Modern mass-produced HEVs prolong the charge on their batteries by capturing kinetic energy via regenerative braking

Recuperator

A **recuperator** is a special purpose counter-flow heat exchanger used to recover waste heat from exhaust gases. In many types of processes, combustion is used to generate heat, and the recuperator serves to recuperate, or reclaim, this heat in order to reuse/recycle it. The term recuperator also refers to compressed air counter-flow heat exchangers used for heat recovery in the highly pressurized air industries. By recovering some of the energy usually lost as waste heat, the recuperator can make an air bearing turbocharger engine or a compressed air turbine significantly more efficient. This recovery of the heat of compressed air significantly increases the efficiency of our air bearing turbine engine.

Rotor

The **rotor** is the non-stationary part of a rotary electric motor or alternator, which rotates because the wires and magnetic field of the motor are arranged so that a torque is developed about the rotor's axis. In some designs, the rotor can act to serve as the motor's armature, across which the input voltage is supplied. The stationary part of an electric motor is the stator. A common problem is called cogging torque.

Seething Virtual State Vacuum Via That Broken Symmetry

Every electrical circuit ever built — and those built today — are in fact powered by electrical energy extracted by the circuit dipolarity from the **local seething vacuum**, from active space itself. But our engineers are trained to build circuits which also self-destroy the extraction of that vacuum energy, faster than they power their loads.

Silicon Energy Storage Battery

The electrolyte of silicon energy adopts silicate composite; making its performance surpass sulfuric acid in all aspects. It eliminates acid fog during manufacturing and deployment. The silicon energy storage battery charges with an internal chemosynthesis method, which is different from the standard lead-acid storage battery.

The following are the operating mechanisms:

1. During the chemosynthesis charging and battery repetitive charging process, the electrolyte added with silicate composite transforms from a solid state to a liquid state. This improves the charging acceptability and lowers the temperature.
2. The above transformation phenomenon provides a better physical condition to overcome the electrolyte stratification. It enables the battery to have a bigger capacity and stronger electric current during charging/discharging.
3. After charging the battery, the silicate electrolyte will not have the silicon split effect. It overcomes the AGM diaphragm and the "islet" effect of the electrode plate. It solves the problem of harm from thermal runaway effect.

Solar Paint

We now have the ability to paint the photovoltaic cells right into the surface of the vehicle. There are epoxy coatings that can be spray painted onto the car that act as an insulator. The solar cells can be spray painted onto the epoxy surface as stripes and connected with small copper jump wires, forming a grid with the proper output to maintain the batteries.

Solar Cells

A **solar cell** or **photovoltaic cell** is a large area electronic[1] device that converts solar energy into electricity by the photovoltaic effect. Photovoltaics is the field of technology and research related to the application of solar cells for solar energy. Sometimes the term *solar cell* is reserved for devices intended specifically to capture energy from sunlight, while the term *photovoltaic cell* is used when the source is unspecified. Assemblies of cells are used to make solar modules, or photovoltaic arrays. Solar cells have many applications. Cells are used for powering small devices such as electronic calculators. Photovoltaic arrays generate a form of renewable electricity, particularly useful in situations where electrical power from the grid is unavailable such as in remote area power systems, Earth-orbiting satellites and space probes, remote radiotelephones and water pumping applications. Photovoltaic electricity is also increasingly deployed in grid-tied electrical systems. Similar devices intended to capture energy from other sources include thermophotovoltaic cells, betavoltaics cells, and optoelectric nuclear batteries.

Stator

The **stator** is the stationary part of an electric generator or electric motor. The non-stationary part on an electric motor is the rotor. Depending on the configuration of a spinning electromotive device the stator may act as the *field magnet*, interacting with the armature to create motion, or it may act as the *armature*, receiving its influence from moving field coils on the rotor. The first DC generators (known as dynamos) and DC motors put the field coils on the stator, and the power generation or motive reaction coils on the rotor. This was necessary because a continuously moving power switch known as the commutator is needed to keep the field correctly aligned across the spinning rotor. The commutator must become larger and more robust as the current increases. The stator of these devices may be either a permanent magnet or an electromagnet. Where the stator is an electromagnet, the coil which energizes it is known as the *field coil* or *field winding*. An AC alternator is able to produce power across multiple high-current power generation coils connected in parallel, eliminating the need for the commutator. Placing the field coils on the rotor allows for an inexpensive slip ring mechanism to transfer high-voltage, low current power to the rotating field coil. It consists of a steel frame enclosing a hollow cylindrical core (made up of laminations of silicon steel). The laminations are to reduce hysteresis and eddy current losses.

Supercapacitors

EEStor is a company based in Cedar Park, Texas, United States that claims to have developed a type of capacitor for electricity storage, which EEStor calls 'Electrical Energy Storage Units' (EESU).[1] ZENN motors proposes to use the EEStor EESU as part of their "ZENNergy" drive-train. Its CEO and president is Richard Weir, who is also a co-inventor named on their principal technology patent. According to its patent application, these units will use high-purity barium titanate coated with aluminum oxide and glass to achieve a level of capacitance claimed to be much higher than what is currently available in the market. The claimed energy density of the prototype is 1 MJ/kg; existing commercial supercapacitors typically have an energy density around 0.02 MJ/kg, while lithium ion batteries are around 0.54–0.72 MJ/kg. For comparison, gasoline has an energy density around 45 MJ/kg. However, an internal combustion engine is unable to extract the majority of this energy and overall energy delivered to the wheel in a motor vehicle is typically less than 20% of this value; a value of 9 MJ/kg is more indicative. EEStor has stated that production examples should be about 2.5 MJ/kg, better than 1/4th the usable energy density of gasoline. This would represent an enormous leap in energy storage technology, which is the reason the company generates so much excitement and skepticism at the same time. (As part of the review process for patents, the U.S. Patent Office does not test the validity of the technical assertions in the patent application. Instead, it only checks whether earlier publications disclose the same idea. Based on these claims, a full charge should give the capacitor sufficient energy to drive a small car 300 miles (480 km). Although the technology should allow very fast charging (e.g., **5 minutes**), standard household wiring is not capable of delivering the power required for this, so charging times this short would probably require purpose-built high capacity dispensing stations. Overnight charging at home should still be practical, as is using a second EESU for the home which could be charged overnight using cheap, off-peak electricity to then charge the EEStor unit in the car in 5-10 minutes on demand - and deliver cheap electric power to the house too, making expensive peak power plants obsolete. Also, according to Ian Clifford, a normal household outlet with 110 volt supply can fully charge the EEStor powered CityZENN in 4 hours for a 250-mile (400 km) range and a normal household outlet with 220 volt supply can fully charge the EEStor powered CityZENN in 2 hours for a 250-mile (400 km) range.

Super High Pressure (SHP)

A magnetic air car vehicle, comprising: a magnetic motor in which permanent magnets are used to turn an output shaft without using 5 electromagnetic components; a first compressor connected to be driven by the magnetic motor and for producing compressed air at a high-pressure (HP) level, about 200-PSI; storage tanks on-board said vehicle for receiving compressed air at said HP level from said first compressor; a series of compressors ducted in totem-pole fashion to multi-stage compress said compressed air from the storage tanks and first compressor beginning with HP level to a very high-pressure (VHP) level, then from said VHP level to an ultra high-pressure (UHP) level, then from said UHP level to a **super high-pressure (SHP),** and finally from said SHP level to an extremely high-pressure (EHP) level of compressed air; a pneumatic torque converter for converting said EHP level of compressed air to drive an output shaft; a transmission/differential connected to receive said output shaft from the pneumatic torque converter, and that includes forward speeds, a neutral, and a reverse; and wheels connected to provide locomotion from the transmission/differential.

Symmetry

Symmetry in physics includes all features of a physical system that exhibit the property of symmetry—that is, under certain transformations, aspects of these systems are "unchanged", according to a particular observation. A **symmetry of a physical system** is a physical or mathematical feature of the system (observed or intrinsic) that is "preserved" under some change.

The transformations may be *continuous* (such as rotation of a circle) or *discrete* (e.g., reflection of a bilaterally symmetric figure, or rotation of a regular polygon). Continuous and discrete transformations give rise to corresponding types of symmetries. Continuous symmetries can be described by Lie groups while discrete symmetries are described by finite groups (see Symmetry group). Symmetries are frequently amenable to mathematical formulation and can be exploited to simplify many problems. An important example of such symmetry is the invariance of the form of physical laws under arbitrary differentiable coordinate transformations.

Thermodynamics

In physics, **thermodynamics** (from the Greek θεϱμη, *therme,* meaning "heat-and δυναμις, *dynamis,* meaning "power") is the study of the conversion of heat energy into different forms of energy (in particular, mechanical, chemical, and electrical energy); different energy conversions into heat energy; and its relation to macroscopic variables such as temperature, pressure, and volume. Its underpinnings, based upon statistical predictions of the collective motion of particles from their microscopic behavior, is the field of statistical thermodynamics, a branch of statistical mechanics. Roughly, heat means "energy in transit" and dynamics relates to "movement"; thus, in essence thermodynamics studies the movement of energy and how energy instills movement. Historically, thermodynamics developed out of need to increase the efficiency of early steam engines.

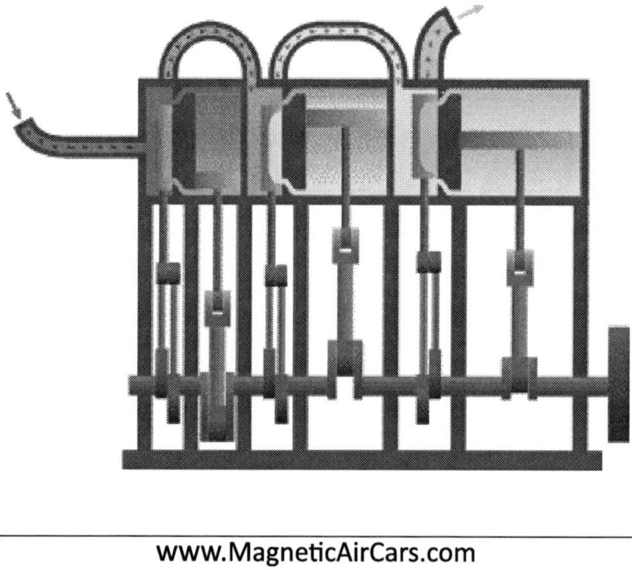

Typical **thermodynamic system**, showing input from a heat source (boiler) on the left and output to a heat sink (condenser) on the right. Work is extracted, in this case by a series of pistons. The starting point for most thermodynamic considerations are the laws of thermodynamics, which postulate that energy can be exchanged between physical systems as heat or work.-They also postulate the existence of a quantity named entropy, which can be defined for any system. In thermodynamics, interactions between large ensembles of objects are studied and categorized. Central to this are the concepts of *system* and *surroundings*. A system is composed of particles, whose average motions define its properties, which in turn are related to one another through equations of state. Properties can be combined to express internal energy and thermodynamic potentials, which are useful for determining conditions for equilibrium and spontaneous processes. With these tools, thermodynamics describes how systems respond to changes in their surroundings. This can be applied to a wide variety of topics in science and engineering, such as engines, phase transitions, chemical reactions, transport phenomena, and even black holes. The results of thermodynamics are essential for other fields of physics and for chemistry, chemical engineering, aerospace engineering, mechanical engineering, cell biology, biomedical engineering, materials science, and economics to name a few.

Torque

Torque is the tendency of a force vector to rotate an object about some axis[1] (or fulcrum or pivot). Just as a force is a push or a pull, a torque can be thought of as a twist. The symbol for torque is τ, the Greek letter *tau*. Torque is also called **moment** or **moment of force**. This should neither be confused with the various other definitions of "moment" in physics nor with "momentum". In the context of mechanical engineering, the terms "moment" and "torque" are not necessarily interchangeable; rather, one or the other may be preferred in a specific context. For example, "torque" is usually used to describe a rotational force down a shaft, for example a turning screw-driver, whereas "moment" is more often used to describe a bending force on a beam. The magnitude of torque depends on three quantities: First, the force applied; second, the length of the *lever arm*[2] connecting the axis to the point of force application; and third, the angle between the two. In symbols:

$$\vec{\tau} = \vec{r} \times \vec{F}$$
$$\tau = rF\sin\theta$$

where
$\vec{\tau}$ is the torque vector and τ is the magnitude of the torque,
$\vec{r}$ is the lever arm vector (vector from the axis to the point of force application), and
r is the length (or magnitude) of the lever arm vector,
$\vec{F}$ is the force vector, and F is the magnitude of the force,
$\times$ denotes the cross product,
θ is the angle between the force vector and the lever arm vector.
The length of the lever arm is particularly important; choosing this length appropriately lies behind the operation of levers, pulleys, gears, and most other simple machines involving a mechanical advantage
The SI unit for torque is the newton meter (N m). In Imperial and U.S. customary units, it is measured in foot pounds (ft·lbf) (also known as 'pound feet') and for smaller measurement of torque: inch pounds (in·lbf) or even inch ounces (in·ozf) .

Tourmaline and Crystals

Tourmaline is a crystal silicate mineral compounded with elements such as aluminium, iron, magnesium, sodium, lithium, or potassium. Tourmaline is classed as a semi-precious stone and the gem comes in a wide variety of colors. The name comes from the Sinhalese word "turamali" or "toramalli", which applied to different gemstones found in Ceylon (now Sri Lanka).

Turbofan Jet Engine

A **turbofan** is a type of aircraft engine consisting of a ducted fan which is powered by a gas turbine. Part of the airstream from the ducted fan passes through the gas turbine core, providing oxygen to burn fuel to create power. However, most of the air flow bypasses the engine core, and is accelerated by the fan blades in much the same manner as a propeller. The combination of thrust produced from the fan and the exhaust from the core is a more efficient process than other jet engine designs, resulting in a comparatively low specific fuel consumption.[1]

A few designs work slightly differently and have the fan blades as a radial extension of an aft-mounted low-pressure turbine unit. Turbofans have a net exhaust speed that is much lower than a turbojet. This makes them much more efficient at subsonic speeds than turbojets, and somewhat more efficient at supersonic speeds up to roughly Mach 1.6, but have also been found to be efficient when used with continuous afterburner at Mach 3 and above. All of the jet engines used in currently manufactured commercial jet aircraft are turbofans. They are used commercially mainly because they are highly efficient and relatively quiet in operation. Turbofans are also used in many military jet aircraft.

Ultra High Pressure (UHP)

A magnetic air car vehicle, comprising: a magnetic motor in which permanent magnets are used to turn an output shaft without using electromagnetic components; a first compressor connected to be driven by the magnetic motor and for producing compressed air at a high-pressure (HP) level, about 200-PSI; storage tanks on-board said vehicle for receiving compressed air at said HP level from said first compressor; a series of compressors ducted in totem-pole fashion to multi-stage compress said compressed air from the storage tanks and first compressor beginning with HP level to a very high-pressure (VHP) level, then from said VHP level to an **ultra high-pressure (UHP) level**, then from said UHP level to a super high-pressure (SHP), and finally from said SHP level to an extremely high-pressure (EHP) level of compressed air; a pneumatic torque converter for converting said EHP level of compressed air to drive an output shaft; a transmission/differential connected to receive said output shaft from the pneumatic torque converter, and that includes forward speeds, a neutral, and a reverse; and wheels connected to provide locomotion from the transmission/differential.

Vacuum

This vacuum means "absence of matter" or "an empty area or space"; for the cleaning appliance, see *vacuum cleaner*. For the vacuum used in defining the MKS system of units, see *free space*. A **vacuum** is a volume of space that is essentially empty of matter, such that its gaseous pressure is much less than atmospheric pressure.[1] The word comes from the Latin term for "empty," but in reality, no volume of space can ever be perfectly empty. A **perfect vacuum** with a gaseous pressure of absolute zero is a philosophical concept that is never observed in practice. Physicists often discuss ideal test results that would occur in a perfect vacuum, which they simply call "vacuum" or "free space" in this context, and use the term **partial vacuum** to refer to real vacuum. The Latin term **in vacuo** is also used to describe an object as being in what would otherwise be a vacuum. The quality of a vacuum refers to how closely it approaches a perfect vacuum. The residual gas pressure is the primary indicator of quality, and is most commonly measured in units called torr, even in metric contexts. Lower pressures indicate higher quality, although other variables must also be taken into account. Quantum theory sets limits for the best possible quality of vacuum, predicting that no volume of space can be perfectly empty. Outer space is a natural high quality vacuum, mostly of much higher quality than can be created artificially with current technology. Low quality artificial vacuums have been used for suction for many years. Vacuum has been a frequent topic of philosophical debate since Ancient Greek times, but was not studied empirically until the 17th century. Evangelista Torricelli produced the first laboratory vacuum in 1643, and other experimental techniques were developed as a result of his theories of atmospheric pressure. Vacuum became a valuable industrial tool in the 20th century with the introduction of incandescent light bulbs and vacuum tubes, and a wide array of vacuum technology has since become available. The recent development of human spaceflight has raised interest in the impact of vacuum on human health, and on life forms in general.

Very High Pressure (VHP)

A magnetic air car vehicle, comprising: a magnetic motor in which permanent magnets are used to turn an output shaft without using 5 electromagnetic components; a first compressor connected to be driven by the magnetic motor and for producing compressed air at a high-pressure (HP) level, about 200-PSI; storage tanks on-board said vehicle for receiving compressed air at said HP level from said first compressor; a series of compressors ducted in totem-pole fashion to multi-stage compress said compressed air from the storage tanks and first compressor beginning with HP level to a **very high-pressure (VHP) level,** then from said VHP level to an ultra high-pressure (UHP) level, then from said UHP level to a super high-pressure (SHP), and finally from said SHP level to an extremely high-pressure (EHP) level of compressed air; a pneumatic torque converter for converting said EHP level of compressed air to drive an output shaft; a transmission/differential connected to receive said output shaft from the pneumatic torque converter, and that includes forward speeds, a neutral, and a reverse; and wheels connected to provide locomotion from the transmission/differential.

Zero Point Energy

In physics, the **zero-point energy** is the lowest possible energy that a quantum mechanical physical system may have and is the energy of the ground state. The quantum mechanical system that encapsulates this energy is the zero-point field. The concept was first proposed by Albert Einstein and Otto Stern in 1913. The term "zero-point energy" is a calque of the German *Nullpunktenergie.* All quantum mechanical systems have a zero point energy. The term arises commonly in reference to the ground state of the quantum harmonic oscillator and its null oscillations. Zero-point energy is sometimes used as a synonym for the vacuum energy, an amount of energy associated with the vacuum of empty space. In cosmology, the vacuum energy is one of possible explanation for the cosmological constant. The variation in zero-point energy as the boundaries of a region of vacuum move leads to the Casimir effect, which is observable in nanoscale devices. A related term is zero-point field, which is the lowest energy state of a field; i.e. its ground state, which is non-zero.

Zero Point Physics

The amount of potential energy contained within the fabric of space has been independently calculated by numerous physicist to be somewhere around the inconceivably high figure of <10 to the 115th power > grams/ cubic centimeter. This means that within a single cubic centimeter of vacuum there is an amount of energy potentially present but in a state of balance of self cancellation equivalent to the energy contained within:

100000000000000000000000
000000000000000000000000
000000000000000000000000
000000000000000000000000
000000000000000000000000
000000000000000000000000

Grams of matter < 115 zeros after 1 >. This is a number that is greater than the estimated number of atoms in the visible universe.

It is a figure that exceeds the estimated amount of energy present in the known or visible universe which includes all the energy being liberated by tens to hundreds of trillions of galaxies all ablaze with trillions of stars. This amount of energy is literally inconceivable by the human imagination. With this amount of energy potentially present in every cubic centimeter of vacuum, it becomes utterly ludicrous to speak of an "energy shortage" as is so often done in today's political and economic spheres. The paradigm of the zero point energy potential of the vacuum of space literally disallows scarcity from being considered to be real.

We have found a way to tap into this sea of energy potential and extract some of it to do work. This is the big issue at hand and the outcome of the project has enormous ramifications for the future existence of the human species and for that of all life on earth.

It is safe to say that the understanding of zero point vacuum fluctuation physics is one of the more important steps that have ever occurred in the history of science. It has the potential to utterly change our view of the nature of things and may have technological ramifications that can allow us to really clean up the planet and stop destroying the biosphere in the greedy, shortsighted way that is currently going on at an ever-accelerating rate. And accomplishing this miracle will set us on the road to an even greater miracle – a fundamental shift in consciousness that can liberate humanity's imagination to the same degree that zero point physics can liberate the energy in so-called empty space